T0231280

*Acram Taji*
*Prakash P. Kumar*
*Prakash Lakshmanan*

# In Vitro Plant Breeding

*Pre-publication*
*REVIEWS,*
*COMMENTARIES,*
*EVALUATIONS . . .*

"**T**his book will guide students to the door of novel plant breeding. After reading this book, you will understand the concepts of in vitro breeding and how these techniques have been developed and applied."

**Takashi Handa, PhD**
*Assistant Professor,*
*University of Tsukuba,*
*Ibaraki, Japan*

"**A**dvances in plant propagation in vitro have been instrumental to the improvement of plants of agricultural and horticultural significance. *In Vitro Plant Breeding* is an excellent introduction to the various methods involved in plant breeding technology. Topics such as micropropagation techniques, haploid plant production, in vitro mutagenesis, somaclonal variation, cryopreservation, in vitro flowering, in vitro tuberization, and molecular plant breeding are covered in depth. Each chapter outlines the principles, methods, and applications of the specific in vitro technique in an easy to understand style.

This comprehensive compilation of various techniques is highly recommended for plant biology undergraduate students as an ancillary textbook in understanding the developments in plant biotechnology. It is useful as an introductory reference book for researchers in small and medium-sized industries engaged in plant genetic improvement. For more experienced researchers, it gives an overview of current applications of in vitro methods."

**V. Sarasan, PhD**
*Research Scientist,*
*Micropropagation Unit,*
*Royal Botanic Gardens,*
*Surrey, United Kingdom*

*More pre-publication*
*REVIEWS, COMMENTARIES, EVALUATIONS . . .*

"*In Vitro Plant Breeding* serves as an easy reference for a wide audience, especially for undergraduate and graduate students pursuing studies in plant biology, molecular biology, plant breeding, biotechnology, conservation biology, and natural resource management. This book encompasses a wide array of subjects by providing useful insights into the current status of plant tissue culture research and its strong applicability in plant improvement and the commercial arena. The interdisciplinary nature of the book also makes it a useful reference for professionals in related fields."

**Chakradhar Akula, PhD**
*Agronomy Department,*
*University of Wisconsin,*
*Madison*

# In Vitro Plant Breeding

# FOOD PRODUCTS PRESS
Crop Science
Amarjit S. Basra, PhD
Senior Editor

New, Recent, and Forthcoming Titles of Related Interest:

*Dictionary of Plant Genetics and Molecular Biology* by Gurbachan S. Miglani

*Advances in Hemp Research* by Paolo Ranalli

*Wheat: Ecology and Physiology of Yield Determination* by Emilio H. Satorre and Gustavo A. Slafer

*Mineral Nutrition of Crops: Fundamental Mechanisms and Implications* by Zdenko Rengel

*Conservation Tillage in U.S. Agriculture: Environmental, Economic, and Policy Issues* by Noel D. Uri

*Cotton Fibers: Developmental Biology, Quality Improvement, and Textile Processing* edited by Amarjit S. Basra

*Heterosis and Hybrid Seed Production in Agronomic Crops* edited by Amarjit S. Basra

*Intensive Cropping: Efficient Use of Water, Nutrients, and Tillage* by S. S. Prihar, P. R. Gajri, D. K. Benbi, and V. K. Arora

*Physiological Bases for Maize Improvement* edited by María E. Otegui and Gustavo A. Slafer

*Plant Growth Regulators in Agriculture and Horticulture: Their Role and Commercial Uses* edited by Amarjit S. Basra

*Crop Responses and Adaptations to Temperature Stress* edited by Amarjit S. Basra

*Barley Science: Recent Advances from Molecular Biology to Agronomy of Yield and Quality* by Gustavo A. Slafer, José Luis Molina-Cano, Roxana Savin, José Luis Araus, and Ignacio Romagoas

*In Vitro Plant Breeding* by Acram Taji, Prakash P. Kumar, and Prakash Lakshmanan

*Bacterial Disease Resistance in Plants: Molecular Biology and Biotechnological Applications* by P. Vidhyasekaran

*Tillage for Sustainable Cropping* by P. R. Gajri, V. K. Arora, and S. S. Prihar

*Plant Viruses As Molecular Pathogens* by Jawid A. A. Khan and Jeanne Dijkstra

*Crop Improvement Challenges in the Twenty-First Century* by Manjit S. Kang

# In Vitro Plant Breeding

Acram Taji
Prakash P. Kumar
Prakash Lakshmanan

**CRC Press**
Taylor & Francis Group
Boca Raton  London  New York

CRC Press is an imprint of the
Taylor & Francis Group, an informa business

Published by

Food Products Press®, an imprint of The Haworth Press, Inc , 10 Alice Street, Binghamton, NY 13904-1580

Reprinted 2009 by CRC Press

Cover design by Anastasia Litwak

### Library of Congress Cataloging-in-Publication Data

Taji, Acram.
   In vitro plant breeding / Acram Taji, Prakash P Kumar, Prakash Lakshmanan
      p. cm.
   Includes bibliographical references (p. )
   ISBN 1-56022-907-1 (alk. paper)—ISBN 1-56022-908-X (soft· alk. paper)
   1 Plant micropropagation. I. Kumar, Prakash P. II Lakshmanan, Prakash III. Title.

SB123.6 .T35 2001
63.5'3—dc21

                                                                        2001016191

# CONTENTS

# ABOUT THE AUTHORS

**Acram Taji, PhD,** teaches crop physiology, horticultural science, and plant biotechnology to undergraduate and graduate students at the School of Rural Science and Natural Resources at the University of New England in Australia. Professor Taji is the recipient of various awards for excellence in teaching and research, including the inaugural Australian Award for University Teaching in 1997, the Australian Society for Plant Physiologists' Prize in 1998, and the highest Japanese Government Senior Research Fellowship for Foreign Specialists in 1997-1998.

**Prakash P. Kumar, PhD,** has taught at the National University of Singapore for the past ten years while supervising various research projects. He teaches plant physiology, plant developmental biology, molecular aspects of plant tissue culture, molecular biology of plant stress, and plant tissue culture and morphogenesis. He is the editor of two international journals, *Plant Cell Reports* and *Plant Cell, Tissue, and Organ Culture.*

**Prakash Lakshmanan, PhD,** is a research scientist (biotechnology) at David North Plant Research Centre in the Bureau of Sugar Experiment Stations in Australia. He taught plant physiology and molecular genetics at the Maharaja Sayajirao University of Baroda for two years. In 1989 he joined a commercial plant biotechnology company in Singapore as a research scientist, and in 1991 he moved to the National University of Singapore. During 1998 he was a senior visiting research scientist in the biotechnology laboratory at the University of New England in Australia.

# Foreword

There are many books now published on the topic of plant tissue culture. *In Vitro Plant Breeding* differs from the others because it provides a particular focus, as the name indicates.

Plant tissue culture offers an array of techniques that complement more conventional plant breeding methods. Micropropagation provides rapid multiplication of newly generated or selected genotypes. Other techniques such as in vitro fertilization and protoplast fusion enable the recombination of genotypes otherwise limited by incompatibility. Conventional breeding can be hastened by the generation of homozygous double haploids via androgenesis or conversely by exploiting increased genetic diversity resulting from somatic variability.

Biotechnology, molecular biology, genetic engineering, and genetically modified organisms (GMOs), are the buzzwords of modern biology. Undoubtedly, the application of molecular biology will have a profound impact on plant improvement during this century. However, there is a tendency to overlook the fact that many applications of plant molecular biology depend on our ability to regenerate whole plants from genetically transformed cells or tissues. The techniques of plant tissue culture will continue to play an important role.

This text will serve multiple purposes. For the undergraduate student of plant science it provides a broad introduction to the scope of plant tissue culture techniques. For the traditional plant breeder or the molecular biologist it provides some insight into the potential and limitations of these in vitro tools.

*In Vitro Plant Breeding* provides sufficient background for the reader unfamiliar with plant tissue culture practice to gain an understanding of the basic principles rather than just a series of recipes. It does not attempt to provide a comprehensive review of the individual topics; this makes the text easy to read. The background information provided is important since, as any practicing tissue culturist will testify, it is frequently necessary to refine the "standard recipe" when initiating new cultures. Many plant species remain recalcitrant in culture. An understanding of the system and the influential factors can be an invaluable guide to the trials needed to overcome such problems.

Numerous other texts provide a more in-depth coverage of some of the individual topics covered here but none provides the same focus on the particular application to in vitro plant breeding. It should be a useful addition to the bookshelf of the plant tissue culturist, the plant breeder, and the plant molecular biologist. Indeed these three groups should be working as a team anyway!

*Richard R. Williams*
*Foundation Professor of Horticulture*
*University of Queensland, Australia*

# Preface

The invention of agriculture, 10,000 years ago, heralded the dawn of civilization. The constant improvements in technology and crop productivity through selection and breeding of plants have contributed to the growth of human civilization. The notable increase in yield and production of various crops, especially the cereals, during the past 50 years through traditional and improvised agricultural techniques leading to the "green revolution" of the 1960s and 1970s is perhaps the most commendable achievement of plant breeding. The research that underpinned this progress has produced huge returns, tripling the world food supply during the past 30 years.

The world population increase has been exponential in the past century and it is projected to reach 8.3 billion by 2020, before stabilizing at around 11 billion toward the end of the twenty-first century.

As the human population increases, the demand for more land area for housing and industrial activities also increases, forcing the conventional agriculture into marginally productive land. Requirements of the continually growing population for food and fiber must be satisfied by an increase in yield. To meet the projected food demand for this increase in population, the average yield of all cereals needs to be increased by 80 percent between now and 2020.

Bearing in mind that there are natural limits to increased productivity by environmental manipulations, one way to improve and maintain a sustainable level of crop productivity is through the exploitation of biotechnology. Indeed, one major aim of biotechnology is to increase yield (biomass), while maintaining stable human ecosystems.

The emergence of plant biotechnology and the search into the regulation of genes that are responsible for specific qualitative and quantitative traits should be able to complement the conventional plant breeding and help overcome the constraints that have limited agricultural productivity.

Plant biotechnology and molecular breeding have already proved their impact in enhancing the productivity of some of the major agricultural crops. They will continue to contribute to the production of plants with novel traits that are otherwise difficult or impossible to develop by conventional breeding.

However, at present there is some concern about the possibility of misuse or adverse effects from this emerging technology. Biotechnology, considered as the technology for the twenty-first century, can be difficult for the general public to understand. Scientists and teachers in this area are concerned that if biotechnology, especially molecular biotechnology, is misunderstood, governments will impose undue strict regulations, hampering progress. Therefore, participation in debates and public education is vital to our further progress in this area.

In an age in which the dynamic field of plant biotechnology advances at an accelerating rate, there is a growing need to explain, in as simple terms as possible, how that science is applied. We believe that many authors fall into the trap of writing books as if all their readers were scientists. In this book, we, as plant scientists with a major interest in plant biotechnology, have aimed to be more embracing and communicate our collective knowledge of 50 years to as diverse an audience as possible. However, in writing the text, we have assumed knowledge of biology and chemistry on behalf of the reader.

For a number of years, the authors have successfully taught undergraduate and postgraduate students and have conducted research in many areas of plant sciences including plant biotechnology. The experience has been useful in writing this book, with students as our target audience. Furthermore, due to our work experience in many parts of the world, the examples used throughout the book are those applicable to researchers and students all over the world.

# Chapter 1

# Introduction

Plant tissue culture is an abbreviation for plant protoplast, cell, tissue, and organ culture. These various types of culture involve, as a common factor, the growth of microbe-free plant material in an aseptic environment such as sterilized nutrient medium in a test tube. In recent years plant tissue culture techniques have developed into a very powerful tool for propagation and breeding of many plant species. The technology began with Gottlieb Haberlandt's speculation regarding cell totipotency at the turn of the twentieth century. Haberlandt suggested that techniques for isolating and culturing plant tissues should be developed and postulated that, if the environment and nutrition of cultured cells were manipulated, those cells would recapitulate the developmental sequences of normal plant growth.

The discovery of auxins, by Went and colleagues, and cytokinins, by Skoog and colleagues, preceded the first success of in vitro culture of plant tissues (Gautheret, 1934; Nobecourt, 1939). White (1943) reported the first successful callus culture of carrots and tobacco. Skoog and Miller (1957) proposed that quantitative interactions between auxins and cytokinins determine the type of growth and morphogenic event that would occur in plants. Their studies with tobacco indicated that a high auxins to cytokinins ratio induced rooting, while the reverse induced shoot morphogenesis. This pattern of response, however, is not universal. While manipulations of auxins to cytokinins ratios have been successful in obtaining morphogenesis in many taxa, it is now clear that many other factors affect the ability of cells in culture to differentiate into roots, shoots, and/or embryos.

A major stimulus for the application of plant tissue culture techniques to the propagation and breeding of many species may be attributed to the early work by Morel (1960) on the propagation of orchids in culture, and to the development and widespread use of a new medium with a high concentration of mineral salts, by Murashige and Skoog in 1962. Since then, this technology has developed considerably and today has a key role in propagation, plant improvement and genetic engineering. Some of the important landmarks in the evolution of in vitro technology are summarized in Table 1.1.

Tissue culture relies on three fundamental abilities of plants:

1. *Totipotency* is the potential or inherent capacity of a plant cell to develop into an entire plant if suitably stimulated. Totipotency implies that all the information necessary for growth and reproduction of the organism is contained in the cell. Although theoretically all plant cells are totipotent, the meristematic cells are best able to express it.

2. *Dedifferentiation* is the capacity of mature cells to return to meristematic condition and development of a new growing point, followed by redifferentiation which is the ability to reorganize into new organs.

3. *Competency* describes the endogenous potential of a given cell or tissue to develop in a particular way. For example, embryogenically competent cells are capable of developing into fully functional embryos. The opposite is noncompetent or morphogenetically incapable.

## *TYPES OF IN VITRO CULTURE*

1. Culture of intact plants (e.g., seed culture in orchids; seedling culture)
2. Embryo culture (e.g., immature embryo culture)
3. Organ culture (e.g., meristem culture)
   - shoot tip culture
   - root culture
   - leaf culture
   - anther culture
4. Callus culture
5. Cell suspension and single cell culture
6. Protoplast culture

TABLE 1.1. Historical Landmarks in the Evolution of In Vitro Technology

- Gautheret (1934), Nobecourt (1937), and White (1934) achieved the first success in developing plant tissue culture, e.g., Gautheret obtained callus formation from cultured explants of tree cambium and phloem tissue.
- After the discovery of cytokinins by Skoog and colleagues, Skoog and Miller in 1957 observed that shoot and root formation are controlled by the auxin/cytokinin balance.
- In vitro somatic embryogenesis was first described by Steward, Mapes, and Mears (1958) and Reinert (1958).
- Anther culture and production of haploid plants was achieved by Guha and Maheshwari (1964; 1966) and Bourgin and Nitsch (1967).
- Protoplast culture, fusion, and development of somatic hybrids were described in the 1960s and 1970s (Cocking 1960; Belliard, Vedel, and Pelletier 1979; Gleba and Sytnik 1984).
- During the 1980s recombinant DNA technology and production of transgenic plants were achieved (Schell 1987; Schell and Vasil 1989).

## APPLICATIONS OF PLANT TISSUE CULTURE

The most common reasons for the application of in vitro techniques to plant production are summarized in Table 1.2, but its most important application in this century will be to crop improvement using gene technology. Therefore, the objective of this book is to introduce the reader to the use of plant tissue culture techniques which can be exploited in plant improvement, and explain how the new technologies may be used effectively in improving the productivity of agricultural crops.

The importance of plants to human beings cannot be overemphasized. We are dependent upon plants for our food, fiber, fuel, medicine, and housing. Therefore, it is not surprising that much human activity has been directed toward improving and producing plants with useful characteristics. Conventional approaches to plant breeding have achieved much. However, these approaches have their limitations. The comparatively recent development in our knowledge of the molecular and cellular mechanisms that underpin the activities and functions of living systems has enabled us to develop novel methods in plant improvement. These techniques focus on the application of molecular and cellular biology. The contribution of plant biotechnology through gene manipulation is not just restricted to increasing yield of crops or generating devices for preventing the ravages of pests and diseases, but extends to helping improve the quality of food and the way we use land. Plant biotechnology, therefore, has considerable potential for growth, and for enhancing the quality of life and well-being of the biosphere. In the chapters that follow we have endeavored to introduce the important techniques used in tissue culture and their usefulness in plant breeding. It is not, however, the intention of this book to provide all of the underpinning science and technology associated with plant tissue culture and biotechnology, but rather the application of plant tissue culture to plant breeding.

TABLE 1.2. Applications of Plant Tissue Culture

---

- Clonal propagation or rapid and large-scale multiplication of genetically identical plants from a single "superior" stock plant.
- Elimination of pathogens. This also facilitates transfer of plant material through international borders.
- Establishment of disease-free in vitro stock plants in culture.
- Germplasm storage and long-term storage of stock plants.
- Selection of mutants from spontaneous or induced mutations.
- Production of rooted micro cuttings in recalcitrant woody ornamental species.
- Recovery of hybrids from incompatible species through either embryo or ovule culture.
- Production of haploid plants through anther culture. Haploid plants may be used to recover recessive mutations in breeding programs. Subsequent regeneration of double haploids provides homozygous and thus pure-breeding lines.

---

# Chapter 2

# Morphogenesis/Organogenesis

## INTRODUCTION

Development, as applied to plants formed by sexual reproduction, refers to the sum of the gradual and progressive qualitative and quantitative changes in growth which comprise the transformation of a zygote into a mature, reproductive plant. The phenomenon is characterized by changes in size and weight, appearance of new structures and functions and losses of former ones. Of course it is quite appropriate to speak of development of individual organs as well as whole plants.

Development can be viewed as a phenomenon consisting of three processes, which normally occur concomitantly—i.e., growth, cellular differentiation, and morphogenesis.

## PLANT GROWTH

Growth is defined as a permanent increase in size. However, size is not the only criterion used to measure growth. For example, the growth of a sample of cells in suspension culture could be assessed by measuring its fresh weight—that is, the weight of the living tissue— at selected time intervals. In other cases, fresh weight may fluctuate due to changes in water status of the plant and so it may be a poor indicator of actual growth. In these situations, measurements of dry weight are often more appropriate.

Cell number is a common and convenient parameter to measure the growth of unicellular organisms, such as the green alga *Chlamydomonas*. In multicellular organisms, cell divisions can occur in the absence of growth, e.g., during the early stages of embryo development the zygote divides into progressively smaller cells with no net

5

increase in the size of the embryo. True growth occurs only when the cells expand.

In general, a reliable way to assess growth is to measure one or more size parameters such as length, height or width and calculate the area or the volume where appropriate. When growth is measured continuously over time, an S-shaped curve like that in Figure 2.1 is often obtained.

Growth curves of this form show a period of slow growth (the lag phase), followed by a period of rapid growth (the logarithmic and linear phases), followed by the period when growth ceases (the stationary phase). Growth curves of this shape may apply to single cells, plant organs, or whole plants. A multitude of factors affect the rate of growth and lead to this type of growth kinetics.

## CELLULAR DIFFERENTIATION

Cellular differentiation is the transformation of apparently identical cells, arising from a common progenitor cell, into diverse cell types with different biochemical, physiological, and structural specializations.

In plants, unlike in most animals, cell differentiation is frequently reversible, particularly when plant tissue is excised and maintained in culture. Such differentiated cells can reinitiate cell division and, given the appropriate nutrients and hormones, even regenerate whole plants. Thus, differentiated cells, with some exceptions, retain all the genetic information (encoded in DNA) required for the development of a complete plant—a property termed totipotency (see Chapter 1). This is because most cells differentiate by regulating gene expression, not by altering their genome (exceptions to this include cells that lose their nuclei, such as phloem sieve tube members, and cells that are dead at maturity, e.g., xylem tracheids).

## MORPHOGENESIS

Morphogenesis refers to the origin of gross form and appearance of new organs, e.g., shoots and roots. It encompasses growth and cellular differentiation, but is a process of a higher order of magnitude which supercedes events occurring in single cells. Interactions among cells have a great effect on the collective fates of individual cells in the developing plant body.

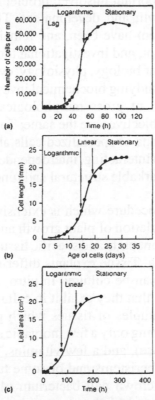

(a)

(b)

(c)

FIGURE 2.1 Measuring plant growth (a) Growth of the unicellular green alga *Chlamydomonas*. Cells are placed in fresh growth medium, under optimum conditions, and growth is assessed by counting the number of cells. An initial lag period, during which cells may synthesise enzymes to metabolize new substrates, is followed by a period of logarithmic or exponential growth during which the rate of increase in cell number increases. Finally, as nutrients and light become limited or the level of toxic substances builds up, the growth rate decreases and the stationary phase is reached. The number of cells may then decrease as cells die. Growth of unicellular organisms under optimum conditions often lacks the linear phase typically observed for multicellular plants (see parts b and c). (b) Growth of the filamentous green alga *Nitella* assessed by measuring cell length. Growth rate increases rapidly (logarithmic phase) and then remains constant (linear phase) for several days. As the cell approaches maturity, growth slows down and stops (stationary phase). (c) Growth of leaves of radish *(Raphanus sativus)*, monitored by measuring leaf area. Again, an S-shaped curve is obtained. Growth is initially slow while cell divisions generate cells that will form the leaf. These cells then expand rapidly (logarithmic and linear phases) until the leaf approaches its mature size and the growth rate decreases (stationary phase). (*Source:* Figure 15 1, p. 374, from *Plant Physiology* by Lincoln Taiz and Eduardo Zeiger. Copyright © 1991 by The Benjamin/Cummings Publishing Company, Inc. Reprinted by permission of Addison Wesley Longman Publishers, Inc.)

One of the most important unsolved problems in all of biology is the regulation of growth and development. The experimental approaches to this problem have been, and remain, many and varied. The knowledge, concept, and investigational tools and techniques of biochemistry, molecular biology, physiology, cytology, and anatomy are all used. Whether studying biochemical investigations of cell-free systems, or the origin of gross morphological features of whole plants, the overlying objectives are the same: To understand how one cell gives rise to millions of specialized cells all of which are partitioned so precisely into interacting, interdependent tissues and organs that they maintain remarkable structural and functional unity as a living system.

An experimental procedure which is extensively employed in investigations of the regulation of plant growth and development is the culturing of excised plant parts and specific tissue explants in vitro on synthetic culture media. Tissue explants differ markedly in the success with which they can be cultured in vitro, and in the degree of morphogenesis, if any, that they exhibit in culture. At one end of the spectrum we have examples of tissues which grow very well on a simple medium containing only a few inorganic salts, a carbon source (for energy, e.g., sucrose), and a few vitamins. Typical examples of these are explants of meristems and of some tumor tissues such as those induced by the crown gall bacterium *(Agrobacterium tumefaciens)*. At the other end of the spectrum are examples of tissue explants, which so far have failed to grow on any "defined medium." In between the two extremes are many examples of explants from nonmeristematic tissues, which require an auxin (Figure 2.2) and a cytokinin for growth in vitro (Figure 2.3). The most common result obtained with tissue explants cultured in vitro is the proliferation of the explant into a mass of relatively undifferentiated and disorganized cells, termed *callus*.

Skoog and colleagues made very important progress in the years after 1947 with explants of pith parenchyma from internodes of tobacco *(Nicotiana tabacum* cv. Wisconsin 38) plants. These explants failed to grow on a basal medium unless supplied with an auxin such as IAA. On media containing auxin the pith explants exhibited cell enlargement, but no cell division or morphogenesis.

| | | | |
|---|---|---|---|
| 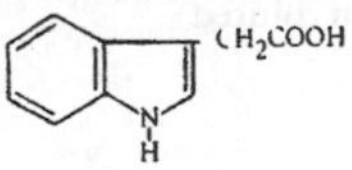 IAA | Indole-3-acetic acid. Sparingly soluble in water. Freely soluble in alcohol. | mol wt. 175 18 |
| 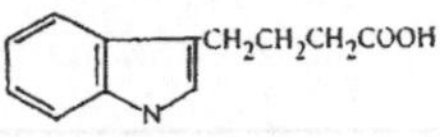 IBA | Indole-3-butyric acid Insoluble in water. Soluble in alcohol | mol.wt. 203.23 |
| 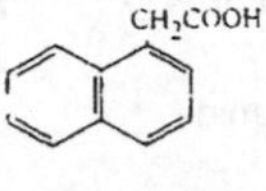 NAA | 1-Naphthaleneacetic acid Insoluble in water Freely soluble in acetone and in ether. | mol.wt 186 20 |
| 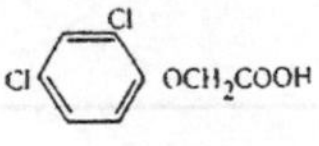 2,4-D | 2-4-Dichlorophenoxyacetic acid. Practically insoluble in water Soluble in acetone and in ether | mol.wt 221.04 |
| 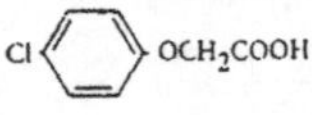 4-CPA | 4-Chlorophenoxyacetic acid Soluble in organic solvents | mol.wt. 186.60 |
| 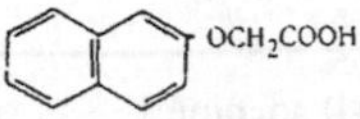 NOA | 2-Naphtoxyacetic acid. Insoluble in water Freely soluble in alcohol | mol wt. 202.20 |
| 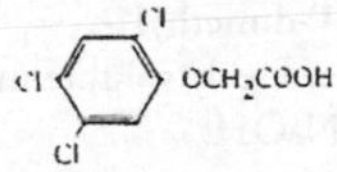 2,4,5-T | 2,4,5-Trichlorophenoxy- acetic acid Soluble in organic solvents | mol wt 255.49 |
| 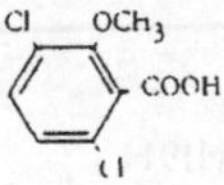 DICAMBA | 3,6-Dichloro-2-methoxybenzoic acid or 3,6-dichloro-o-anisic acid. Soluble in ethanol and acetone. | mol wt 221 04 |
| 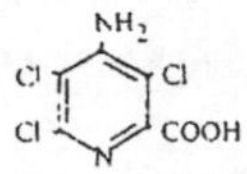 PICLORAM | 4-Amino-3,5,6-trichloro-2- pyridinecarboxylic acid or 4-amino-3,5,6-trichloropicolinic acid Soluble in water | mol wt. 241.48 |
| 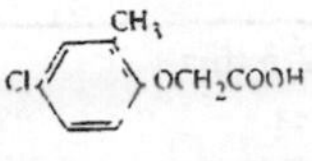 PCPA | 2-methyl-4-chlorophenoxy- acetic acid As sodium salt soluble in water | mol wt 241 48 |

FIGURE 2.2. Auxin and auxin-like compounds used in tissue culture.

| | | | |
|---|---|---|---|
| 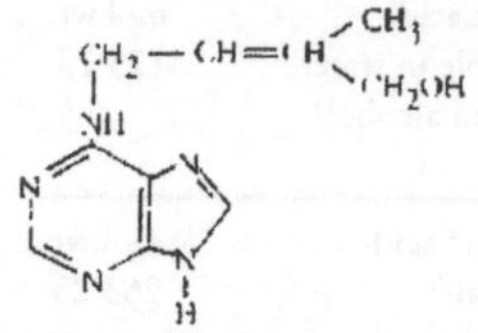 | ZEATIN | 4-hydroxy-3-methyl-trans-2-butenylaminopurine<br>Freely soluble in diluted HCl or NaOH. | mol.wt.<br>219.25 |
| 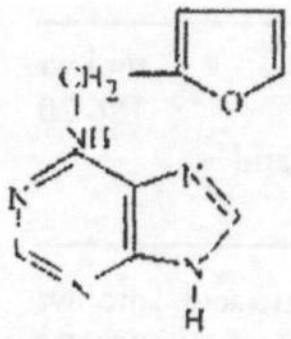 | KINETIN | 6-furfurlaminopurin or $N^6$-furfuryladenine.<br>Freely soluble in diluted HCl or NaOH. | mol.wt.<br>215.21 |
| 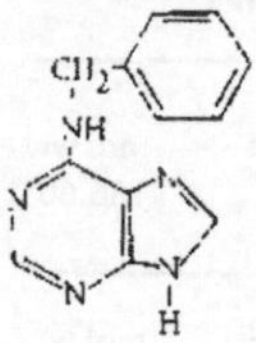 | BAP | 6-benzylaminopurine.<br>Freely soluble in diluted HCl or NaOH. | mol.wt.<br>225.26 |
| 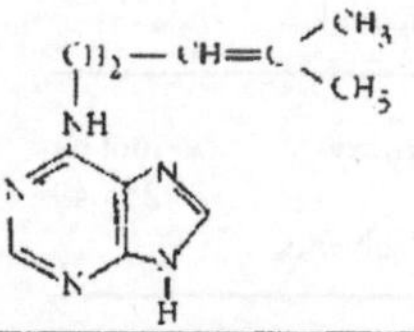 | ZIP | $N^6$ (2-isopentyl) adenine or $N^6$ ($\Delta^2$-isopentenyl) adenine or 6-($\gamma\gamma$-dimethyl)-allylaminopurine. Freely soluble in diluted HCl or NaOH. | mol.wt.<br>203.25 |
| 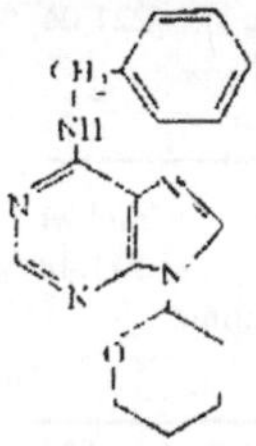 | PBA | 6-(benzylamino)-9-(2-tetrahydropyranyl)9H purine | mol.wt.<br>309.37 |
| 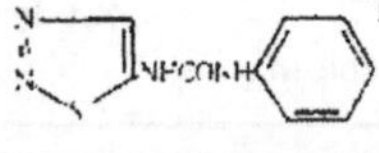 | THIDIAZURON | 1-phenyl-3-(1,2,3 thia-diazol-5-yl) urea.<br>Soluble in water. | mol.wt.<br>220.25 |

FIGURE 2.3. Some synthetic and naturally occurring cytokinins as well as cytokinin-like compounds used in tissue culture.

Skoog and colleagues discovered that if vascular tissue was placed in contact with the pith explants cultured in vitro on a medium containing auxin, the tissue was stimulated to undergo cell division. A long search for the chemical stimulus supplied by the vascular tissue began. Surprisingly, autoclaved herring sperm DNA added to the nutrient medium had a powerful cell division-promoting effect. A single chemical substance was isolated from the heat-treated DNA that would, in the presence of auxin, stimulate tobacco pith parenchyma tissue to proliferate in culture. This substance was identified as 6. furfurylamino purine and named kinetin (Miller et al., 1955).

Kinetin stimulated cell division in tobacco pith when it was cultured on a medium containing an auxin. It is important to note that kinetin is not a naturally occurring plant growth regulator and it does not occur as a base in the DNA of any species. It is a by-product of the heat-induced degradation of the DNA in which the deoxyribose sugar of adenosine is converted to a furfuryl ring and shifted from the 9 to the 6 position on the adenine ring.

The discovery of kinetin was important because it demonstrated that cell division could be induced by a simple chemical substance. More important, when a synthetic molecule initiates a biological response, frequently it does so because the synthetic molecule has many of the same properties as naturally occurring molecules that regulate that response in the organism. Thus, the discovery of kinetin suggested that naturally occurring molecules with a structure similar to kinetin may regulate cell division activity within the plant. Of course, this proved to be the case. Several naturally occurring cytokinins from various microbial, plant, and animal sources have since been isolated (Figure 2.3). The cytokinins are now firmly established as a major group of plant hormones which participate in the normal regulation of growth and development.

Skoog and Miller (1957) described a unique example of hormonal regulation of morphogenesis in tobacco pith explants. This important research elegantly documented a delicate and quantitative interaction between auxin and cytokinin in the control of bud and root formation from pith explants With a particular combination of concentrations of IAA and kinetin, the pith tissue grew as relatively undifferentiated callus. By varying the ratio of IAA:kinetin, however, they could successfully cause the explants to give rise to buds or roots (Figure 2.4).

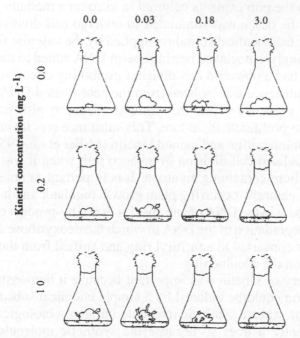

FIGURE 2.4. Cells of most dicots can be grown in culture if provided with auxin, cytokinin, some vitamins, minerals, and sugar. The ratio of auxin to cytokinin is important, as are the absolute concentrations. If both are too low, there is no growth or development. With high auxin and low cytokinin (upper right), roots form. With low auxin and high cytokinin (lower left), buds form in some species. With relatively equal amounts, the tissue grows as an irregular mass (callus) containing patches of xylem.

Thus, by varying the amounts of the two types of growth substances in the culture medium, morphogenesis in tobacco pith explants can be controlled to a remarkable degree. The implications for this kind of research for undertaking normal, as well as abnormal or pathological growth are great indeed.

Comparatively few generalizations can be derived from tissue culture studies, but the auxin:cytokinin ratio often seems to be important in differentiation of callus to form either organized tissues or specialized cells such as xylem. Most studies consider only the effects of these two plant growth regulators (PGRs, Figure 2.5). Other factors are undoubtedly involved, and the work with tobacco callus was extended to show that the ratio of gibberellins to cytokinin controlled

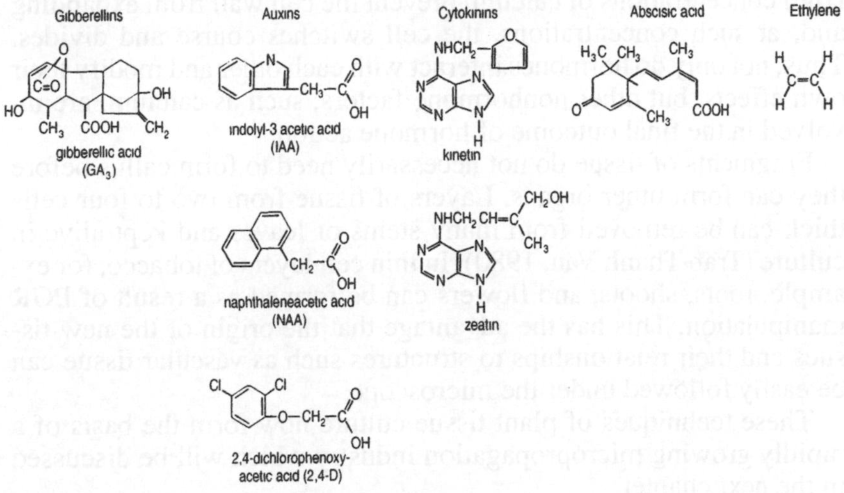

FIGURE 2.5. Structural formulas of some plant growth regulators used in tissue culture media.

both the number of shoots formed and their shape (in the presence of gibberellic acid (GA₃) there are fewer shoots and they have elongated leaves).

Another point is that PGRs are not the only regulators involved. In lilac and bean callus, both xylem and phloem elements are formed at appropriate ratios of auxin to cytokinin. The relative amounts of the two types of vascular tissue depend not on PGR concentration but on the concentration of sucrose in the medium. For example, in lilac culture the concentration of 1.5 to 2.5 percent sucrose causes xylem formation, while 4 percent sucrose favours phloem formation. Based on this, it is certainly tempting to suggest that a high sugar level stimulates the development of the cell type (phloem) that is responsible for its transport.

The following example illustrates further the complexity of involvement of auxin and cytokinin in the control of morphogenesis.

In a study using tuber tissue of the Jerusalem artichoke (*Helianthus tuberosus*), it was shown that a third substance—the calcium ion—can modify the action of the auxin-cytokinin combination. In this study, IAA plus low concentrations of kinetin was shown to favor cell enlargement, but as $Ca^{+2}$ was added to the culture, a steady shift occurred in the growth pattern from cell enlargement to cell division. High concentrations of calcium prevent the cell wall from expanding and, at such concentrations, the cell switches course and divides. Thus, not only do hormones interact with each other and modify their own effects, but other nonhormone factors, such as calcium, are involved in the final outcome of hormone action.

Fragments of tissue do not necessarily need to form callus before they can form other organs. Layers of tissue from two to four cells thick can be removed from many stems or leaves and kept alive in culture (Tran Thanh Van, 1980). In thin cell layers of tobacco, for example, roots, shoots, and flowers can be formed as a result of PGR manipulation. This has the advantage that the origin of the new tissues and their relationships to structures such as vascular tissue can be easily followed under the microscope. .

These techniques of plant tissue culture now form the basis of a rapidly growing micropropagation industry which will be discussed in the next chapter.

# Chapter 3

# Micropropagation

## DEFINITION

Micropropagation is the true-to-type propagation of a selected genotype using in vitro techniques. Several stages are involved in micropropagation, each of which is influenced by an array of physical, nutritional, and hormonal factors. Micropropagation only makes sense when adequate starting material is used. Therefore, the choice of stock material cannot be made indiscriminately. For most ornamentals and fruit crops the starting material is an elite plant, selected for a certain phenotypic characteristic, e.g., the flavor of the fruit, the color of the flower, the shape of the bush, etc. In forestry and vegetable production, the elite material will either be a selected phenotype or a plant with elite seed which will be used as starting material for a more or less extensive micropropagation program.

Many commercial operations, especially those propagating ornamental plants, follow the desirable procedure of making a regular positive mass selection in the micropropagated material that is grown to maturity in the greenhouse. This selected material is then reintroduced in culture. This method avoids many problems and even has the intrinsic value of improving the selection.

## STAGES IN MICROPROPAGATION

There are five stages in micropropagation (de Fossard, 1976): The preparative stage (Stage 0); the initiation stage (Stage 1); the multiplication stage (Stage 2); the rooting stage (Stage 3); and the transplanting stage (Stage 4), which are described in detail.

## Stage 0: The Preparative Stage

Initially, this stage was introduced to overcome the huge problem of contamination. Raising stock plants in a greenhouse under more hygienic conditions can considerably reduce the risk of contamination, especially those related to fungal infestation. However, it is more difficult to interpret the results with respect to bacterial contamination since most often we cannot distinguish between endogenous and exogenous bacteria.

### Growing Stock Plants Under More Hygienic Conditions

Ideally, stock plants should be grown in a greenhouse. Not only will this reduce the population of microorganisms that are living on the surface of the plant, but also it will help to produce quality plants for Stage 1 due to regular application of fertilizer, fungicide, and insecticide.

### Changing the Physiological Status of the Stock Plant, the Source of Explants, or the Explant

Stage 0 also includes many interventions that can make an explant more suitable or more reliable as starting material. The most commonly manipulated parameters are light, temperature, and plant growth regulators.

*Light.* Controlling the photoperiod in a greenhouse opens possibilities of yielding more standardized explants throughout the year. This is especially true for those plants where flowering is under photoperiodic control; e.g., in vitro culture of *Begonia* sp. was positively influenced not only by high temperatures but also by long days applied to stock plants.

The work of Paul Read (1988) in the United States demonstrated the effect of light quality on stock plant management and subsequently on performance of plants in vitro. He showed that if petunia plants were given a period of red light (640-700 nm) treatment at the end of the day, they branched profusely, whereas those with far-red (700-795 nm) light treatment grew upright and remained unbranched. Furthermore, leaf segment cultures initiated from stock plants treated with red light produced up to three times as many shoots per explant as did explants taken from nontreated plants.

*Temperature.* The importance of temperature in affecting physiological changes that improve the efficiency of micropropagation has been demonstrated in several plant species.

In most bulb crops and temperate trees there is a need for cold to break dormancy. For example, storage of lilies at 4°C had a strong effect on bulblet production in vitro. Furthermore, the weight of the bulblets was influenced by temperature. In hyacinth, storage of cultures for 70 days at 15°C produced heavier bulblets than those stored for the same length at 25°C.

When stock plants of some woody plants were kept at cool temperatures of 4 to 5°C for a period of time, equivalent to that required in nature to break dormancy, additional flushes of new growth were produced.

*Pretreatment with plant growth regulators.* The reactivity of an explant in Stage 1 can be controlled by an appropriate treatment with plant growth regulators of the stock plant, the source of explants, or the explant itself. However, limited research has been carried out on the effect of pretreatment with plant growth regulators (both promoters and retardants) on the response of explants in vitro. By manipulating the stock plants we can change their physiology and subsequently their response in vitro. Maene and Debergh (1986), in the University of Gent in Belgium, injected benzyladenine (BA) solution (up to 500 milligrams per liter) into the trunk of *Magnolia soulangana* to improve success at Stage 1. Similarly, germination of *Brassica campestris* seedlings in the presence of BA was found to enhance the subsequent regeneration efficiency of cotyledon explants.

Another approach is to put the source of explants in a forcing solution, e.g., solution of sucrose, 8-hydroxyquinoline citrate containing BA and gibberellic acid ($GA_3$). This improves the responsiveness of woody plants in Stage 1.

Pulse treatment of the primary explants of *Grevillea* spp. in solutions containing cytokinin has resulted in a proliferation rate equivalent to that attained by culturing the explants on medium supplemented with cytokinins.

### Stage 1: Initiation Stage

The purpose of this stage is to produce axenic cultures. Factors influencing the success at this stage are:

## The Explant

For most micropropagation work the explant of choice is an apical or axillary bud (Figure 3.1); for only a limited number of plants, other explants are used, e.g., leaf pieces in *Begonia* and *Saintpaulia* (African violet) or flower heads in *Gerbera* spp.

These factors determine the success rate in stage one.

- The age of the stock plant
- The physiological age of the explant
- The developmental stage of the explant
- Size of the explant

## Hypersensitivity Reactions

When plant tissues are exposed to stress situations such as mechanical injury (as is the case with isolation of an explant from the stock plant), metabolism of phenolic compounds is stimulated. This intervention leads to hypersensitivity reactions such as:

- The release of content of broken cells
- Reactions in the neighboring cells, but without showing symptoms of injury themselves
- And/or the premature death of specific cells in the environment of the wound or the place of infection

In general, the metabolism of phenolic compounds has three possible types of reactions in response to stress or injury:

1. Oxidation of preformed phenolic components (giving rise to quinones and polymerized material)
2. Synthesis of monomeric derivatives
3. Synthesis of polymeric phenolic derivatives

The synthesis of monomeric phenolics in the undamaged tissues can lead to the accumulation of large quantities of preformed products, or to the appearance of new products that play a role in the protection mechanism of the tissue. The role of these products can be to form a physical barrier (lignin) against invasions or an inhibitor of microbial growth (quinones, phytoalexins).

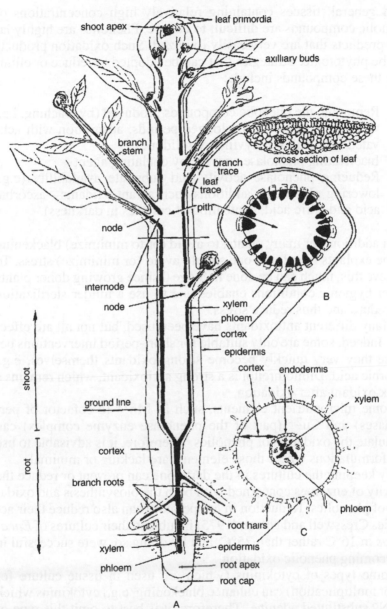

FIGURE 3.1. *A*, the principle organs and tissues of the body of a seed plant; *B*, cross-section of stem; *C*, cross-section of root. Apical or axillary buds are good sources of explants.

In general, tissues containing relatively high concentrations of phenolic compounds are difficult to culture. Phenolics are highly labile products that are very easily oxidized. Such oxidation products can be phytotoxic. Strategies that can be adopted to reduce or eliminate these compounds include:

• Removing the phenolic compounds produced (by leaching, i.e., washing in water for prolonged periods, adsorption with activated charcoal or polyvinylpyrrolidone)
• Inactivating phenolase enzyme by chelating agent
• Reducing phenolase activity and substrate availability (e.g., lowering pH, media additions such as antioxidants—ascorbic acid and citric acid, incubation of cultures in darkness)

In addition, for many plants, to avoid (or to minimize) blackening of the explants in culture, one should avoid (or minimize) stress. To achieve this, much can be done in Stage 0 since growing donor plants under hygienic conditions enables one to use a milder sterilization procedure and thus reduce stress.

Many different antioxidants have been used, but not all are effective. Indeed, some are only suitable for short-period interventions because they very quickly become strong oxidants themselves, e.g., ascorbic acid. Dithiothreitol is a strong antioxidant, which remains a weak oxidant after oxidation.

Some micronutrient elements, such as $Mn^{2+}$ (a cofactor of peroxidases) and $Cu^{2+}$ (part of the phenolase enzyme complex) can stimulate the oxidation of phenolics. Therefore, it is advisable to use salt formulations where those elements are lacking or minimal.

By keeping the cultures in the dark, one can prevent or reduce the activity of enzymes concerned with both the biosynthesis and oxidation of phenolics. Reduction in temperature can also reduce their activities Cresswell and Nitsch (1975) incubated their cultures of *Eucalyptus* in 16°C rather than 25°C, and by doing so, were successful in overcoming phenolic oxidation.

Some types of cytokinins (which are used in tissue culture for shoot multiplication) can enhance blackening, e.g., cytokinins which are $N^6$ substituted adenine. Therefore, it is best to omit this type of cytokinin when initiating cultures. In general, use of any plant growth regulators at initiation stage should be avoided or minimized.

At high salt concentration, phenolic compounds that are produced in the plant cannot leach out. The possibility of an osmotic effect preventing leaching cannot be excluded.

## Stage 2: Multiplication

In this stage, axenic cultures that have been produced from Stage 1 are placed on a cytokinin-rich medium to produce numerous shoots. This stage can be repeated a few cycles until an adequate number of shoots is established for the rooting stage (Stage 3).

During this stage, explants may either produce callus-forming cultures (callogenesis) or shoot-forming cultures (caulogenesis).

In the case of callus, it may produce embryoids and then each embryoid may give rise to a new plantlet (somatic embryogenesis, see Figure 3.2), or it may produce meristemoids which then grow to a shoot (organogenesis). Of course, production of callus and taking the micropropagated plant through this route is not desirable unless we

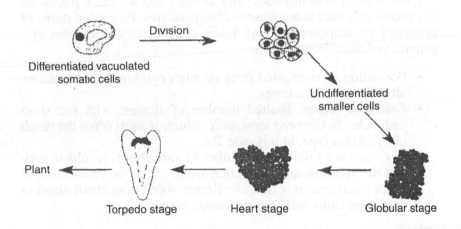

FIGURE 3.2 Through somatic embryogenesis a somatic (non-gametic) cell undergoes differentiation to form a bipolar structure containing both root and shoot axes. Given the right physical and chemical conditions, these somatic embryos can mature and germinate. *(Source:* Smith, 1992, p. 57.)

are specifically interested in somatic embryos for artificial seed production. Callogenesis often leads to genetic aberrations known as somaclonal variation (see also Chapter 9), and therefore the resultant plants may not be true-to-type (off-types).

The second pathway, depending on the source of explant, is as follows: if the source of explant is anything else but the apical or axillary buds (e.g., root segments, leaf segments, petiole segments), as with *Saintpaulia* and *Begonia rex,* the shoots obtained are called adventitious shoots. Although this method is quick and necessary for multiplication of certain plants, it does produce some off-types.

The best and safest method is axillary caulogenesis in which the starting material is either a terminal bud or an axillary bud (Figure 3.3). This method entails the fewest genetic abnormalities and can therefore be used to produce clonal plants with minimal risks.

The susceptibility and, as a consequence, the reaction of an explant changes with the time in culture and/or the number of subcultures (Debergh and Maene, 1981). After many subcultures, plants which originally produced axillary shoots may only produce adventitious shoots. This means that the way of multiplication in a given system is not necessarily fixed.

Even when a system yields only axillary shoots, these plants are not necessarily safe from genetic abnormalities. Excessive doses of cytokinin or inappropriate cytokinin can be responsible for epigenetic variation. For example:

- Formation of variegated form on high cytokinin concentration after several subcultures.
- Excessive leaves, limited number of flowers with too short peduncles in *Gerbera jamesonii,* which is most often the result of cytokinin type, in this case BA.
- In *Fragaria* an unlimited number of subcultures results in serious disorders in the production fields, such as absence of, or weak rhizogenesis, excessive flower formation, small-sized or deformed fruits and heterogeneous plants.

However, one of the major physiological, anatomical, and morphological disorders encountered at this stage is the problem of hyperhydration (previously known as vitrification). This disorder, which is mainly mainfested in the leaves, affects photosynthesis and gaseous exchange in affected plants. Up to 60 percent of hyperhydrated plants do not acclimatize to outside environment.

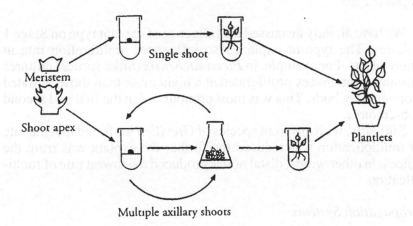

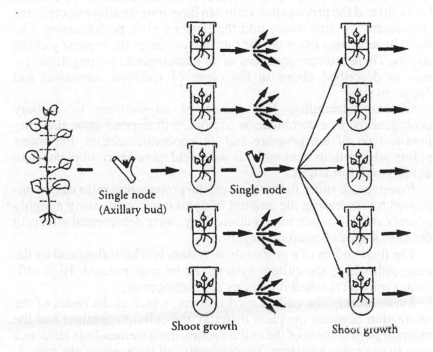

FIGURE 3.3. Axillary caulogenesis can be achieved by using shoot apex or axillary bud as starting plant material.

*Explant Type*

We have already discussed the influence of explant type on Stage 1 cultures. The type of explant also influences proliferation rate in many plants. For example, in *Picea sitchensis* (Sitka spruce) cultures initiated from apices proliferated at a higher rate than those initiated from axillary buds. This was most pronounced in the first and second sub-cultures.

Similarly, with different species of *Grevillea* we found that the rate of multiplication was reduced the further the explant was from the apices. In other words, distal nodes produced the lowest rate of multiplication.

*Propagation Systems*

The choice of the propagation technique, adventitious or axillary (or both) and the propagation ratio can have long-lasting effects in the subsequent in vitro stages and the later ex vitro performance. The best choice is very often not the one that produces the highest yield of shoots. There are consequences of an inappropriate propagation system as described above in the cases of *Gerbera jamesonii* and *Fragaria* sp.

*Genetic variability.* The choice of adventitious or axillary caulogenesis, or a combination of both, will depend upon the trueness-to-type of the progeny and on economic factors. For many plants, adventitious caulogenesis will yield more aberrant plants than axillary bud formation.

*Propagation ratio.* In many cases, the propagation ratio can be improved by increasing the amount of cytokinins or by using a highly potent cytokinin; such interventions may cause detrimental effects in the later stages of micropropagation.

The final choice of a propagation system will be influenced by the ease with which the culture system can be manipulated. High efficiency may be reached if cultures are homogenous.

*Environment.* The culture environment, which is the result of the interaction between the plant material, the culture container and the external environment of the culture room, has a tremendous influence on a tissue culture system. Theoretically, all parameters are considered to be the same in all culture containers in the same culture room and, as a consequence, this also holds true for those from different culture rooms. Therefore, the objective is to provide uniform conditions for all culture containers.

Choice of container lids for culture vessels that affect $CO_2$, water vapor and ethylene gas concentration should be considered carefully. The headspace gas (air space within the culture vessels) can have significant effect on shoot regeneration as shown by Kumar, Reid, and Thorpe (1987) working with *Pinus radiata* cotyledon cultures. A similar observation has been made in tissue culture of several other species. Therefore, selection of a culture container that permits gas exchange with minimum moisture loss would be beneficial in optimizing regeneration for most of the micropropagation systems.

### Stage 3: Pretransplant (Rooting Stage)

The purpose of the third stage is to prepare the plantlet for transplanting from the artificial heterotrophic environment of the test tube to an autotrophic free-living existence in the greenhouse and to their ultimate location. This preparation may involve rooting, but it also involves a change in the physiology of the plantlet so as to stimulate photosynthesis, nutrient and water absorption through roots, and developing resistance to desiccation and to pathogens. At this stage, propagation can be handled in three basic ways:

1. Individual cuttings can be made from a cluster of shoots developing in a culture container and planted directly into a rooting medium in mist or under high humidity with or without rooting hormones. All remnants of the agar medium should be washed off the individual cuttings to remove the potential source of contamination.
2. Individual propagules may be recultured into new containers in a sterile medium with reduced or omitted cytokinin, an increased auxin concentration, and often a reduced inorganic salt content. For some kinds of plants, adventitious roots then develop readily in the propagules; better root hair development may occur in an aerated liquid medium than in an agar medium. With other plants, rooting is best if the propagule is kept in the auxin medium for only one or two days, and then transferred to an auxin-free medium (root induction is promoted by auxin but root growth could be inhibited by the presence of auxin). Or, the

propagule may simply be dipped into a rooting (auxin) solution immediately and inserted directly into an auxin-free medium.
3. The third method is to include an "elongation" phase between Stage 2 and Stage 3 by placing the propagules into an agar medium for two to four weeks without cytokinin (or at very low levels) and, in some cases, adding (or increasing) gibberellic acid. The propagules are then handled as described in the previous methods.

Selection for uniform propagules and discarding of the inferior and abnormal plantlets should occur at the start of Stage 3. Sometimes, plant cultures deteriorate with time, lose leaves, fail to grow, develop tip-burn, go dormant, or lose potentiality to regenerate. Plants of species having an inherent dormancy or rest requirement may need to be chilled to stimulate new growth and elongation.

### Stage 4: Transplanting

Rooted or unrooted plantlets are removed from the culture vessel, agar is washed away completely to remove a potential source of contamination, and the plantlets are transplanted into a standard pasteurized soil mix in small pots in a relatively conventional manner. Initially, plantlets should be protected from desiccation in a shaded, high humidity tent or under mist. Several days may be required for new functional roots to form. Temperatures similar to the growth room and light intensity (30 percent of the ambient light) are also critical factors. Nutrients of the potting mix comprise another limiting factor. The rule of thumb is that no nutrient is provided in the potting mix while plantlets are under the mister (i.e., three to four weeks after transplant).

Once the plantlets are established in the potting medium, they should gradually be exposed to a lower humidity and a higher light intensity. Any dormancy or resting condition that develops in the plant may need to be overcome as part of the establishment process.

### COMMERCIAL MICROPROPAGATION

The technology of plant tissue culture for mass propagation is well developed for industrial applications. There are some 500 commercial micropropagation laboratories worldwide, and most of them have been established in conjunction with big nurseries to satisfy in-house needs for planting material. Most notable commercial success has been in the production of orchids, ornamental, and foliage plants. Several laboratories are not associated with their own nurseries and propagate plants under contract for other growers and may provide additional services, i.e., disease indexing and germplasm storage. To expand micropropagation to new areas, such as plantation, forest, and vegetable crops, the cost of micropropagation should be reduced significantly. Cost studies indicate that the labor input is a significant percentage of total production costs (40 to 70 percent).

Research and development (R&D) in micropropagation laboratories is focused on improvement of micropropagation techniques to reduce propagation costs. Techniques for reducing contamination, new biological systems of micropropagation (e.g., somatic embryogenesis), and automation of the tissue culture system using robots continue to be developed.

## Examples of Commercial Micropropagation of Ornamental Plants

### Group I

Group I includes species cultured easily through mass production. A large number of these plants are free from known pathogens, sustain year-round production, high quality, improved habits and other desirable horticultural characteristics. Examples of this group are *Alstomeria, Anthurium, Caladium, Chrysanthemum, Diffenbachia, Drosera, Gerbera,* gloxinia, *Gypsophila, Heliconia, Freesia, Musa* (banana), *Nepeta, Nephrolepis, Philodendron, Rhododendron, Rosa, Santpaulia,* etc.

### Group II

Plants in this group are amenable for in vitro culture, pathogen tested, produce superior clones, breeding stocks, and germplasm collection. Examples of this group are *Begonia, Dianthus, Gladiolus, Haemanthus, Hemerocallis, Hosta, Hyacinth, Iris, Lilium, Pelargonium* (virus-tested stocks), *Petunia* (hybrids and male sterile lines), etc.

## Group III

Cultured with some difficulty, R&D is required of these plants to improve in vitro systems to assure high quality products. Examples of this group are mainly woody ornamentals such as *Acer, Chamaecyparis, Juniperus, Paeonia, Potentilla, Sequoia, Taxus,* and some of the ornamental palms e.g., *Howeia* and many Australian native plant species e.g., *Grevillea.*

## Examples of Commercial Micropropagation of Small Fruits and Grapevine

### Strawberry

Millions of tissue cultured strawberry plants are being produced annually around the world. Meristem culture combined with thermotherapy is used for virus elimination. Tissue culture is used for increasing the number of rare lines, i.e., virus-indexed stocks and new clones.

### Raspberries and Blackberries

Raspberries and blackberries are propagated by single-node cutting. There are well established protocols for rapid propagation of disease-free plants and new cultivars ( e.g., thornless blackberries).

### Blueberries

Blueberries are propagated readily from cuttings. Tissue culture propagation can be used for multiplication of superior plants, which will be used to make cuttings.

### Kiwifruit (Actinidia)

Limited material from kiwifruit can be propagated by shoot tip culture.

### Grapes

Several in vitro techniques are available for grape propagation and virus elimination. Tissue culture can be accepted as a tool for grape stock production (disease free, selected and local clones, new hybrids, and cultivars). Both scions and rootstocks can be propagated in vitro. Potential exists for new table grape propagation.

### Examples of Fruit Trees Suitable for Commercial Micropropagation

#### Apple (Malus)

Tissue culture of apple is used mainly for propagation of rootstocks. The field performance of micropropagated scion cultivars must be carefully observed before commercial propagation.

#### Cherries (Prunus)

Protocols are available for both sweet and sour cherries. Micropropagation is used for selected rootstocks.

#### Peaches and Apricots (Prunus spp.)

Only a limited number of peach cultivars and rootstocks have been propagated in vitro. Very little has been reported on micropropagation of apricots. Difficulties and inconsistencies in rooting in both species are the main obstacles to utilizing micropropagation on a commercial scale.

#### Pear (Pyrus spp.)

The in vitro system is not well developed for commercial propagation of the pear.

In any commercial setting a close linkage between micropropagation and breeding program is desirable. Micropropagation should aim at producing:

- Virus-indexed stocks
- New cultivars
- Difficult to propagate elite genotypes
- Large quantities of rootstocks

To establish efficient protocols for local genotypes R&D should be addressed as a matter of priority.

### Other Crops Available for Commercial Micropropagation

#### Asparagus

Superior strains of asparagus are being micropropagated successfully.

#### Vegetables (Cucumber, Pumpkin, Tomato, Onion, etc.)

Micropropagation is used only for specific genotypes, e.g., male sterile parent lines needed for $F_1$ seed production.

#### Garlic

Production of virus-indexed material and germplasm conservation has been achieved in garlic.

#### Spices and Food Flavoring Plants

There are limited applications for mass propagation of spices and food flavoring plants since the market for these products is relatively small. The potential for breeding programs is, however, high.

#### Medicinal Plants (Catharanthus, Digitalis, Solanum laciniatum, etc.)

Protocols are available only for micropropagtion systems of medicinal plants. However, further attention in R&D is required to develop reliable cell culture systems for drug production.

#### Rare and Endangered Plants

In vitro mass propagation can be used for the recovery of endangered plants and for the increase of species populations. Limited possibilities exist for commercialization of resultant plants.

#### Tropical Root and Tuber Crops

Examples of tropical root and tuber crops are shown in Table 3.1.

#### Tropical Fruit Trees

Examples for successful tissue culture propagation of tropical fruit trees are shown in Table 3.2.

TABLE 3.1. Tissue Culture of Tropical Root and Tuber Crops

| Crops | Explants | Results |
|---|---|---|
| Cassava *(Manihot esculenta)* | Meristem tip | Virus-indexed plants |
| Coco yam *(Xanthosoma caracu)* | Shoot tip | Multiple shoots |
| *(X brasiliense)* | Shoot tip | Virus-indexed plants |
| *(X. sagittifolium)* | Shoot tip | Virus-indexed plants |
| Hausa potato *(Coleus parviflorus)* | Leaf | Multiple shoots |
| Sweet potato *(Ipomoea batatas)* | Meristem tip | Virus-indexed plants |
| Sweet yam *(Amorphophallus)* | Corm segments | Shoots |
| Taro *(Colocasia esculenta)* | Shoot tip | Virus-indexed plants and multiple shoots |
| Yam *(Dioscorea alata)* | Shoot tip | Multiple shoots |
| | Bulbil segments | Multiple shoots |
| Yam *(Dioscorea microstachya)* | Nodal segments | Multiple shoots |

TABLE 3.2. Tissue Culture Propagation of Tropical Fruit Trees

| Plant Family | Species | Common Name |
|---|---|---|
| Anacardiaceae | *Mangifera indica* | Mango |
| Annonaceae | *Annona cherimola* | Atemoya |
| | *Annona squamosa* | Custard apple |
| Bromeliaceae | *Ananas comosus* | Pineapple |
| Caricaceae | *Carica papaya* | Papaya |
| | *Carica heilbornii* | Babaco |
| Moraceae | *Ficus carica* | Fig |
| | *Morus alba* | Mulberry |
| | *Morus. indica* | Mulberry |
| Musaceae | *Musa* sp. AAA group | Banana |
| | *Musa* sp. AAB group | Plantain |
| | *Musa* sp. ABB group | Bluggoe |
| Myrtaceae | *Eugenia jambos* | Rose apple |
| | *Eugeni. malaccensis* | Malay apple |
| | *Myrciaria cauliflora* | Jaboticaba |
| Passifloraceae | *Passiflora edulis* | Passion fruit |
| Rosaceae | *Eriobotrya japonica* | Loquat |

TABLE 3.2 (continued)

| Plant Family | Species | Common Name |
|---|---|---|
| Rutaceae | Citrus aurantiifolia | Lime |
| | Citrus aurantium | Sour orange |
| | Citrus grandis | Pomelo |
| | Citrus limettioides | Sweet lime |
| | Citrus limon | Lemon |
| | Citrus medica | Citron |
| | Citrus paradisi | Grapefruit |
| | Citrus reticulata | Mandarin |
| | Citrus sinensis | Sweet orange |
| | Poncirus trifoliata | Trifoliate orange |
| Solanaceae | Solanum quitoense | Naranjilla or Lulo |

## APPLICATIONS OF MICROPROPAGATION

### Pathogen Elimination

Superficial contaminants are eradicated by surface sterilization of explants. To survive and grow in vitro, tissue cultures need to be free of fungal and bacterial pathogens which can infect plants systematically. In vitro axenic practices eliminate most fungal, bacterial, nematode and insect pests. However, viruses, viroids, and certain mycoplasmas can be cultured and multiplied within an in vitro plant.

Most virus diseases are not transmitted through seeds. Seeds of infected plants normally develop into healthy plants. However, sexually developed seeds cannot be used for maintenance of clones of vegetatively propagated plants.

### Methods of Virus Elimination

1. *Heat treatment (thermotherapy)* is the exposure of plants to elevated temperatures, and is used for the elimination of viruses or mycoplasmas. It may be used alone or in conjunction with chemotherapy (Figure 3.4), meristem culture or fragmented shoot apices.
2. *Use of anti-viral agents (chemotherapy)* is the use of chemicals to pretreat diseased source plants prior to excision. These chem-

icals may be incorporated into media to support some therapeutic objective, for example, malachite green or viraxole (ribavirin) is used to eliminate virus in meristem culture.

3. *Meristem culture* consists of meristematic dome tissue and usually one pair of leaf primordia. They are excised for virus elimination. Virus elimination using this technique followed by a rapid clonal propagation is now commonly used for ornamental plants such as *Chrysanthemum,* carnations, etc., and in some crop plants, such as potatoes and cassava, in strawberries, selected fruit trees, and in grapevine.
4. *Fragmented shoot apices* are widely used in the propagation of grapevine.
5. *In vitro micrografting.* (Discussed in Chapter 11.)

*Virus Indexing*

Material originating from in vitro propagation should be considered as "virus-tested" and not "virus-free" as often indicated. Various techniques such as indicator plants, electron microscopy, nucleic acid analysis, and serological tests such as enzyme-linked immunosorbent assay (ELISA) are routinely used for virus indexing. ELISA is a sensitive serological test used for detection and quantification of viruses, proteins, and small molecules such as hormones. Performed on a microtiter plate, many samples can be tested at the same time both rapidly and economically. The most widely used form of ELISA testing uses the double antibody sandwich which involves the addition of a specific antibody to the test plate where it adsorbs, followed by addition of the test sample. Specific particles in the sample are immobilized on the antibody film. Subsequently, enzyme-labeled antibody is added and becomes immobilized on the sample particles. Test particles are then quantified by the addition of enzyme substrate, through the colorometric or fluorometric detection of the reaction product.

### Tissue Culture for Maintenance of Plant Genetic Resources (Germplasm Storage)

The conservation of plant genetic resources is important to ensure future access to valuable genes for plant breeding programs. Germ-

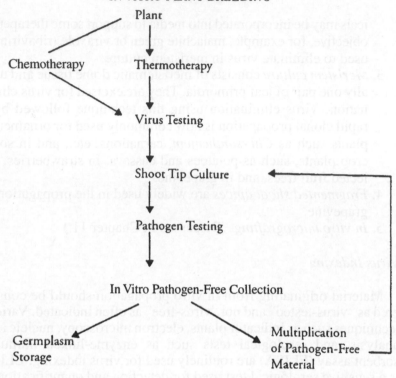

FIGURE 3.4. Combination of thermo-(chemo) therapy with shoot tip culture used for virus elimination.

plasm may be stored in situ in nature reserves or on farm collections. Ex situ forms of storage include seed banks, field gene banks, and in vitro gene banks. For each species or gene pool, an integrated conservation strategy needs to be developed and this may involve a number of these approaches.

In vitro storage techniques have been developed over the past 10 to 20 years and are now firmly established as an alternative for those species for which ex situ storage as seed is not a viable option (Withers and Engelmann, 1995). This includes crops that are vegetatively-propagated (e.g., potato, yam, cassava) or species with recalcitrant seed (e.g., coconut, mango).

In vitro techniques may be utilized at various stages in the conservation and use of plant germplasm. The most obvious application is for the storage of germplasm as slow-growth cultures or as cryo-preserved tissue which can be regenerated using tissue culture techniques. Other applications include the use of tissue culture during field collection of samples (Withers, 1995), virus elimination from valuable germplasm, the international exchange of germplasm as sterile tissue culture samples, and the rapid propagation of germplasm for breeding programs or other end-use applications (Figure 3.5).

## Types of In Vitro Storage

### Storage Using Slow Growth Techniques

Slow growth regimes are used as a medium-term storage option. These techniques enable subculture intervals to be extended to between one and four years for many species, which dramatically reduces maintenance costs.

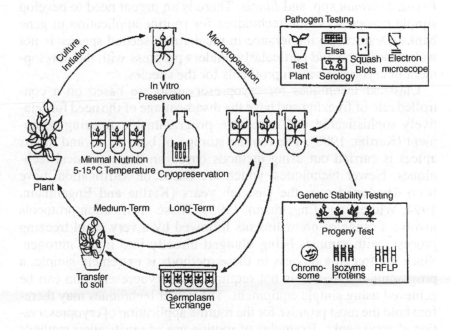

FIGURE 3.5. Techniques used in germplasm storage.

Techniques for slow growth include reduced temperature, reduced light conditions, modifications in media (particularly the addition of osmotic inhibitors or growth retardants), dehydration of tissue, or modifications to gaseous environment (Withers and Engelmann, 1995).

## Storage Using Cryopreservation

Cryopreservation of plant material is the only viable option for the long-term storage of germplasm of species which cannot be stored as seed. At the ultra-low temperatures used for cryopreservation (–196 °C, temperature of liquid nitrogen), cell division and metabolic processes stop, thus plant material can be stored without modification or alteration for unlimited periods of time.

At present, cryopreservation protocols are still in the development stage for most crop species. The few crops for which cryopreservation is routinely used across a range of genotypes include *Rubus, Pryus, Solanum* spp. and *Elaeis*. There is an urgent need to develop simple cryopreservation techniques for routine application in gene banks. Desiccation intolerance in recalcitrant seeded species is not well understood and particularly hinders progress with the development of cryopreservation protocols for the species.

Classical techniques for cryopreservation are based on a controlled rate of freezing and have the disadvantage of the need for relatively sophisticated and expensive programmable freezing equipment (Kartha, 1985). The routine storage of both *Pyrus* and *Rubus* apices is carried out using methods based on these classical techniques. Newer techniques, which are based on vitrification, have been developed over the past ten years (Kartha and Engelmann, 1994; Withers and Engelmann, 1995). These vitrification protocols involve a range of pretreatments followed by a very rapid freezing process, with samples being plunged directly into liquid nitrogen. Since the freezing process in these methods is extremely simple, a programmable freezer is not required and cryopreservation can be achieved using simple equipment. The newer techniques may therefore hold the most promise for the routine application of cryopreservation at gene banks. Examples of routine use of vitrification methods across a range of genotypes or accessions are pregrowth desiccation used for 80 accessions of *Elaeis guineensis* (Dumet, 1994), and the droplet freezing method applied to 150 accessions of *Solanum* spp. (Schafer-Menuhr, Schumacher, and Mix-Wagner, 1994).

## Application of Micropropagation Techniques in Mutation Breeding

### The Search for Genetic Variation

There are two types of variations—qualitatively and quantitatively inherited variations. Because of their all-or-nothing effects, qualitatively inherited traits are easiest ones to deal with. The search for quantitatively inherited variations needs much more attention. First, it is important to recognize whether or not the variation is a consequence of environmental or genetic factors. Second, these variations are not all or nothing effects, but are smaller quantitative differences. Thus, careful quantification is essential.

The important sources for plants exhibiting genetic variability are:

- Existing modern varieties
- Spontaneous or induced mutants
- Primitive races/uncultivated material from centers of origin
- Uncultivated species preserved and maintained in gene banks

The production of mutants for use in in vitro plant breeding programs is described in Figure 3.6.

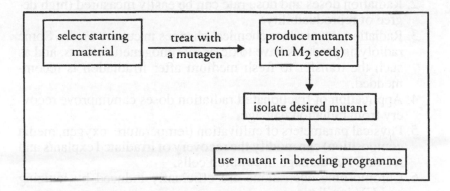

FIGURE 3.6. Scheme for production of mutants for use in in vitro plant breeding programs. (M = plant treated with mutagens.)

*Mutagen Application in the In Vitro System*

One of the limitations to in vivo plant improvement is the dependence on naturally occurring variation. The artificial induction of mutation by radiation (for example X ray, gamma irradiation, thermal neutron and ultraviolet (UV) light) and by chemical means (for example N-ethyl-N-nitrosurea (NEU), N-methyl-N-nitro-N-nitroso-guanidine (MNNG), and ethyl-methane-sulphonate (EMS)) provides a method for creating changes that cannot occur naturally. Some of these techniques are described as follows. It is, however, important to note that apart from the type of mutagen and the dose of mutagen that is applied, factors such as the plant species, the ploidy level of plant or tissue and conditions during the treatment are also important in the success of in vivo plant improvement.

1. Before culture: treatment of whole plants or vegetative propagula (cuttings, bulbs, buds etc.), explants are prepared from $M_1$ material (chimeric).
2. During in vitro culture mutagens can be applied to: shoot tips, explants, adventitious buds on original explants, proliferating culture (multiple axillary shoots), single node cuttings, etc.

*Physical Mutagens (Ionizing Radiation, UV)*
This group of mutagens have the following characteristics:

1. High penetration in multicellular systems, excluding UV.
2. Radiation doses and dose rate can be easily measured (high degree of reproducibility).
3. Radiation can produce chemical changes in culture media. Some radiolysis products have toxic or morphogenetic effects, and as such the transfer to fresh medium after irradiation is recommended.
4. Application of fractionated radiation doses can improve recovery from radiation damage.
5. Physical parameters of cultivation (temperature, oxygen, media composition) can modify the recovery of irradiated explants and repair DNA damage in cultured cells.
6. Chromosome aberrations are efficiently induced by ionizing and UV irradiations.
7. Thin layers or monolayers of cells should be exposed to UV radiation in open containers under aseptic conditions.

## *Chemical Mutagens (Alkylating Agents Are Frequently Used)*

1. The following parameters should be considered when using chemical mutagens:
   - Dose (concentration × time)
   - pH
   - Physical and chemical properties of the agent ("half-life")
   - When applied in vitro, mutagen solutions should be filter-sterilized
   - Interactions with the culture medium
   - Posttreatment conditions.
2. Penetration is difficult in a multicellular system. Carrier agents, e.g., dimethyl sulfoxide (DMSO) can increase uptake of a mutagen.
3. Difficulties in reproductibility may be overcome by standardizing application methods.
4. Since most of the chemical mutagens are carcinogens, the experiments have to be performed in a biohazard box.

## *Chimeras*

Chimera refers to the coexistence of cells of more than one genotype within a genet (Figure 3.7). Chimerism in shoot tips is a commonly occurring phenomenon after mutagen treatment. However, in vitro culture techniques can help to overcome chimerism, which has to be considered especially in in vitro mutagenesis in vegetatively propagated plants.

1. Mosaicism: instability in the progeny of the cell, due to chromosomal, genetic or cytoplasmic mutations.
2. Chimerism: character of a whole tissue, organ or plant with cell lines differing in their caryological or genetic constitutions.
3. Type of chimeras:
   - Solid for first genotype of cells
   - Sectorial chimera
   - Solid for second genotype of cells
   - Periclinal chimera
   - Mericlinal chimera.

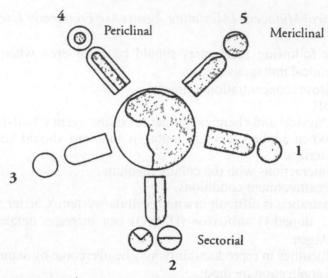

FIGURE 3.7. Plant chimeras contain genetically different tissue. (*Source:* Adapted from Jones, 1937.)

## Mutation

The changes in a gene's internal structure, which give rise to new alleles, are known as mutations. Mutations are changes in genetic information (nucleotides of DNA) and are ultimately responsible for the inherent variation in all living organisms. They may be the product of replacing one type of nucleotide with another or a product of the removal (deletion) of nucleotides, or the addition of extra nucleotides. Mutations can operate at a single nucleotide level (i.e., at an individual gene level) or may involve parts of chromosomes, whole chromosomes (aneuploidy), or even whole genomes (polyploidy).

An important feature of mutations is that by changing the nucleotides, the information stored in DNA is also changed. Depending on the nature of these changes this may give rise to changes in the phenotypic traits of the plant. Mutation occurs naturally (spontaneously) or can be induced in a variety of ways. The position of a mutated cell is important for the realization of a mutation. Mutagenic treatment of an apical meristem will give rise to a mericlinal chimera (Figure 3.8). Periclinal structures can be obtained with repeated in vi-

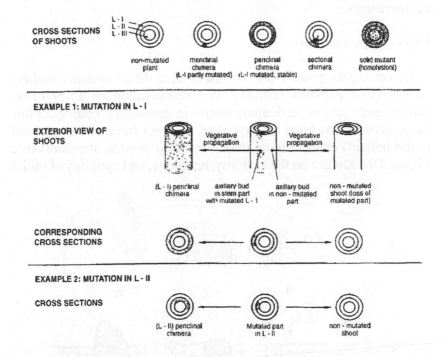

CROSS SECTIONS OF SHOOTS

L - I
L - II
L - III

non-mutated plant

mericlinal chimera (L-I partly mutated)

periclinal chimera (L-I mutated, stable)

sectonal chimera

solid mutant (homohistont)

EXAMPLE 1: MUTATION IN L - I

EXTERIOR VIEW OF SHOOTS

Vegetative propagation

Vegetative propagation

(L - I) periclinal chimera

axillary bud in stem part with mutated L - I

axillary bud in non - mutated part

non - mutated shoot (loss of mutated part)

CORRESPONDING CROSS SECTIONS

EXAMPLE 2: MUTATION IN L - II

CROSS SECTIONS

(L - II) periclinal chimera

Mutated part in L - II

non - mutated shoot

FIGURE 3.8. Chimerism in shoots as observed at some distance from the apical region. In examples 1 and 2 it is shown how, starting from a mericlinal chimera (L-I or L-II partly mutated), after vegetative propagation, either a stable periclinal chimera or a completely nonmutated shoot is obtained. In example 2 the mutation may remain unobserved, depending on the genetic character involved. (*Source:* Broertjes and Van Harten, 1978, p. 23.)

tro propagation of shoot derived from the treated apex either through axillary proliferation or single node culture.

## Competition

Competition may exist between a mutated cell and the nonmutated surrounding tissue, which may eventually lead to the elimination of the mutated cells. Loss of chimerism can be controlled by manipulation of culture conditions. Factors such as culture media, plant growth regulators, and temperature play an important role. Also, the rapid rate of growth in vitro may cause an alteration in the chimeric structure of apical meristems.

## Direct Organogenesis

Direct organogenesis (in rare cases also direct somatic embryogenesis) on explants resulting in a regeneration of adventitious shoots (embryos) is particularly useful in mutation breeding. In this case, adventitious buds (embryos) are formed from one single cell (solid mutant) or from more than one cell (chimeric structure) (see Figure 3.9). Studies on the anatomy, histology, and cytology of initial

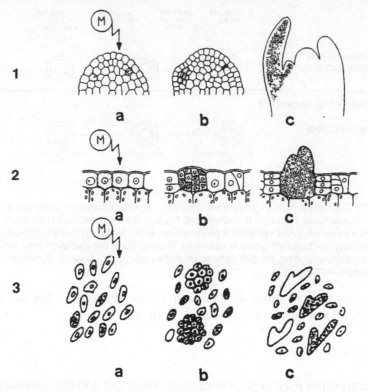

FIGURE 3.9. Origin of mutated structures after in vitro mutagenesis: (1) Chimeric structure of apical meristem; (2) "Solid" adventitious bud originates from single epidermal cell; (3) "Solid" somatic embryos originate from single cultured cells. (M = mutagen treatment.)

explants and on their behavior in vitro is important in understanding the development of mutated sectors (shoots). Figure 3.10 summarizes mutation systems as applied to vegetatively propagated plants.

## Micropropagation Techniques

Micropropagation techniques (as described earlier) can be employed in specific cases for in vitro selection and for multiplication of selected material for field evaluation in different ecological conditions. Micropropagation is convenient for clonal multiplication of new cultivars and their introduction to agricultural practice. Figure 3.11 shows the stages involved in mutation breeding.

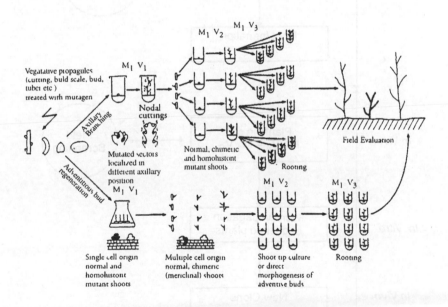

FIGURE 3.10. Schematic representation of in vitro mutation system applied for vegetatively propagated plants. (M = Plants treated with mutagent, in different generations; V = Steps of vegetative propagation.)

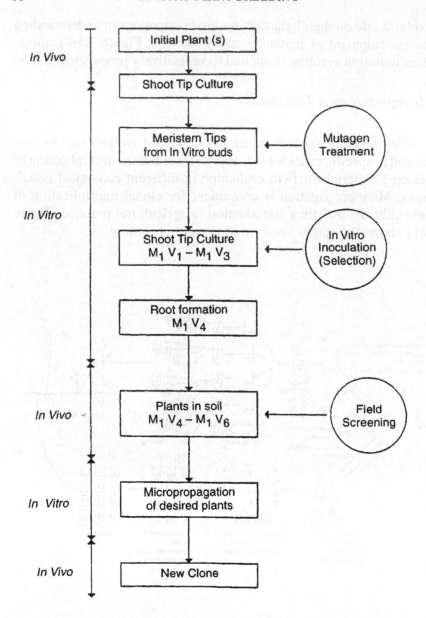

FIGURE 3.11. Proposed in vitro mutation breeding technology for vegetatively propagated plants. (M = Plants treated with mutagent, in different generations; V = Steps of vegetative propagation.)

# Chapter 4

# Haploid Plant Production In Vitro

The chromosome number in a gamete is referred to as the haploid complement, whereas the plant that produces the pollen (male gametes are produced inside the pollen) and the egg (which is the female gamete) has double the haploid number and is referred to as diploid. In many plant species whole plants can be generated from gametes (either a male gamete (androgenesis) or a female gamete (gynogenesis)) under specific culture conditions in vitro. These plants will have a haploid number of chromosomes, that is, they contain one set of genes in their genome.

## *ANATOMY OF ANTHER*

The stamen is a highly specialized structure. It shows clear differentiation into a sterile stalk (filament) bearing the somewhat swollen anther at the tip. The number of stamens in a flower can vary from one to several hundred—either all maturing at the same time or in a graded sequence—commonly from the periphery to the center (centripetal), but occasionally from the center to the periphery (centrifugal, e.g., in *Hibbertia*).

Anthers of the majority of angiosperms are made of four sporangia (tetrasporangia, see Figure 4.1) organized around a sterile mass of tissue (the connective tissue). However, there are bisporangiate anthers, for example in *Ricinus communis* (castor bean), and octosporangiate anther in *Bixa* sp. In avocado *(Persea americana)* and cucumber *(Cucumis sativus)* both bisporangiate and tetrasporangiate anthers may be found in the same flower.

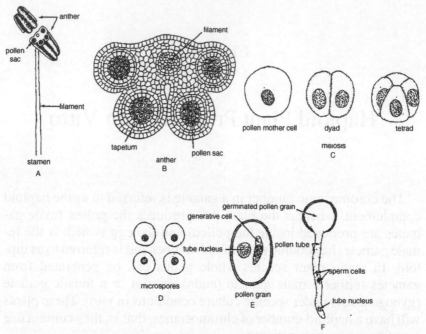

FIGURE 4.1. Development of pollen from a pollen mother cell to the pollen grain. *A*, stamen; *B*, cross-section of anther; *C*, development of tetrad of cells from the pollen mother cell by meiosis; *D*, four microspores; *E*, pollen grain; *F*, germination of pollen grain into pollen tube.

## Derivation of the Anther Wall

Very young anthers are characterized by the presence of a homogenous mass of cells bound by a well-defined epidermis. As it grows, it becomes four-lobed and in each of these lobes a row of cells immediately beneath the epidermis becomes distinguished by larger nuclei and well-stained cytoplasm. These cells are called the archesporial cells.

Each archesporial cell divides in a plane parallel to the surface (periclinal division) resulting in a primary parietal cell towards the outside, and primary sporogenous cell towards the inside.

The primary parietal cell then undergoes a series of divisions, the number depending on species, and 2 to 14 cell layers are formed beneath the epidermis. The epidermis itself remains single-layered throughout but many undergo anticlinal (at the right angles to the surface) divisions and considerable expansion to accommodate the increased amount of tissue underneath. However, in some plants only remnants of the epidermis are left in the mature anther. It is important to note that the epidermis is hairy in some genera (e.g., *Calycanthus*) and bear stomata in several others.

## Composition of Anther Wall

- *Endothecium*—Helps in dehiscence by a longitudinal slit in species where the anther has no valve or pore.
- *Middle layer*—Considered as evolutionary relics, but may be used as a source of food reserve in some species.
- *Tapetum*—This is the innermost layer cells of the anther wall. Its maximum development corresponds with completion of meiosis in pollen mother cells and only traces of it are left in a mature anther. Tapetum is responsible for nutrition of the developing microspore and also for laying down the exine on the pollen. Exine is the outer wall of the pollen grain and is made of the toughest substance of plant origin known as sporopollenin (an alcohol which is also found in the walls of spores and is related to suberin and cutin but much more durable, resulting in spores and pollen grains surviving for millions of years). The inner wall of pollen grain is called intine and is made of cellulose.

## ANTHER CULTURE

As described earlier in this chapter, the cells of haploid plants contain a single complete set of chromosomes and these plants are useful in plant breeding programs for the selection of desirable characteristics.

There are several ways by which haploid plants can be produced in vivo:

- *Gynogenesis:* development from eggs penetrated by sperm but not fertilized (ovules).
- *Androgenesis:* the development of a haploid individual from a pollen grain.
- *Genome elimination:* arises as a result of certain crosses; fertilization occurs but soon afterward some genome is eliminated.

- *Semigamy:* in this case, the nucleus of the egg cell and the generative nucleus of the germinated pollen grain divide independently, resulting in a haploid chimera (a plant whose tissues are of two different genotypes).
- *Chemical treatment:* chromosome elimination can be induced by the addition of certain chemicals, e.g., chloramphenicol.
- *Shocks:* with high or low temperature.
- *Irradiation:* with x ray or UV light.

This chapter covers androgenesis, which is the production of haploid and double haploid plants originating from very young pollen cells. In androgenesis, which only takes place in vitro, the vegetative or generative nucleus of a pollen grain is stimulated to develop into a haploid individual without undergoing fertilization.

The history of androgenesis goes back to the 1960s when Guha and Maheshwari (1964) obtained haploid plants in *Datura* anther culture, followed by Bourgin and Nitsch (1967) who obtained fully grown haploid plants from two species of *Nicotiana, N. sylvestus,* and *N. tabacum.*

Success of androgenesis involves two important considerations. These are:

1. the physiological state of the donor plant, and
2. the conditions necessary to modify the normal development of the pollen towards embryogenesis.

This embryogenesis may be

1. *direct,* i.e., an embryo differentiates directly from the pollen grain (microspore), or
2. *indirect,* i.e., first a callus develops from the pollen grain and then an embryo or an adventitious shoot regenerates. This type of development is not favored since usually a mixture of ploidy levels is produced.

## Condition of Donor Plants

To allow for the in vitro development of pollen into an adult plant, it is very important to start with healthy pollen cells. Plant breeders know from practice that one pollen cell is not the same as another;

good pollen comes from good healthy plants. This is even more important for in vitro culture than for in vivo plant breeding. Therefore, it is only from plants grown under the best growing conditions that one can obtain suitable pollen.

1. The donor plant should be taken care of from the time of flower induction to the sampling of pollen. They should be well fed and placed in an optimum light regime. Artificial light should be avoided as much as possible, for example, plants grown in growth chambers under artificial light conditions may not yield androgenic pollen.
2. Stress of any kind (nutrition and water) should be avoided at all stages of plant growth, particularly during flower induction and over the period of pollen development.
3. Application of any kind of pesticide (externally or systemic) or fungicide for three to four weeks prior to sampling should be avoided. The reason for this is that when cultured in vitro the tissue is more easily killed by chemicals and usually does not recover from such treatments.
4. The state of pollen development when the flower is cut from the donor plant is critical for success in androgenesis. Success is heavily dependent on the variety. Therefore, it is necessary before starting any in vitro culture to follow the development of the pollen from the tetrad stage to the binucleate stage by staining, e.g. using the Feulgen technique or acetocarmine (Prakash, 2000). This technique allows for differential staining of the two nuclei—the vegetative nucleus stains pale pink while the generative nucleus stains dark red.

As there is variation in each cultivar, the state of pollen development suitable for androgenesis goes from just before the first pollen mitosis until just after (see Figure 4.2). However, the best stage is always before the appearance of starch grains. In cereals, when the microspore is formed, its nucleus sits next to the pore. Just before pollen mitosis, the nucleus migrates to the side opposite the pore and mitosis occurs (resulting in generative and vegetative cell production).

Next, the vegetative nucleus migrates back towards the pore and the generative cell is at the side opposite the pore. This cell then di-

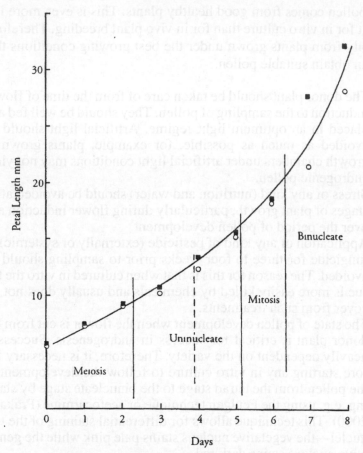

FIGURE 4.2. Relationship between petal length and anther stage in *N. tabacum* 'White Burley.' (*Source:* Sunderland and Wicks, 1971, p. 215.)

vides to form two gametes. Haploid plants can be obtained if the pollen is put into culture when the nucleus is in the center of the grain, i.e., halfway in its migration. In Solanaceae, depending on the variety, the pollen must be taken either before mitosis (e.g., *Nicotiana alala*), during mitosis *(N. tabacum),* or just after mitosis *(N. sylvestris).* It is important that we recognize the critical stage for any species under study (Figure 4.2) and to bear in mind that not all pollen cells are in the same developmental stage. This explains why only small portions of all grains turn into embryoids in any given culture.

# ANDROGENESIS

The underlying principle of androgenesis is to stop development of the pollen cell whose fate is normally to become a gamete, i.e., a sexual cell, and to force its development directly into a plant, as is done with somatic cells. This "abnormal" pathway is possible if the pollen cell is taken away from its normal environment in the living plant and placed in other specific conditions.

## Conditions Favoring Androgenesis

Low temperature enhances the induction of androgenesis. In addition to the synchronizing effect of the cold, it also stops existing metabolism. If the cold period is long enough, it may allow the resting pollen cell to start out on a non-metabolic developmental phase when placed in culture. The cold treatment has the effect that, at the first division of the microspore, two identical nuclei are formed rather than one vegetative and one generative nucleus. The effect of cold treatment can be compared to a sort of induced dormancy or vernalization. The degree of cold applied is dependent on the species. In the case of Solanaceae a cold treatment of not less than three days at 5°C is recommended. It is important to note that anthers of these plants (tobacco, tomato, eggplant, etc.) can last up to two weeks in cold without injury to the pollen.

Culture medium used for pollen and anther culture is another important factor.

1. The major elements of Murashige and Skoog (MS) medium are important in induction of androgenesis. However, some adjustment to the concentration of $NH_4+$ may be required.
2. The microelements and vitamins do not seem to be critical for the induction of androgenesis. However, these become more important during the development of the young embryo into a plant, e.g., chelated iron salt has been shown to play an important role for the differentiation of globular embryos into heart-shaped embryos and further on to complete plants.
3. Sugar is a constituent of anther culture medium, however, its concentration is species specific. In general, 2 to 3 percent sucrose seems to be optimal.

4. No plant growth regulators are needed for induction of androgenesis. If anything, addition of plant growth regulators can be
detrimental or cause complications, such as callusing, which is
highly undesirable at this stage.

As mentioned earlier, androgenesis is a mechanism by which pollen cell development is modified and forced toward embryogenesis,
thus avoiding the sexual pathway. Growth substances inhibit embryogenesis. Enhancing cell division of the vegetative pollen cell while it
is still enclosed in the exine gives a good start toward embryogenesis.
However, the presence of plant growth regulators, especially auxins
and cytokinins in the culture medium, increases the cell division to
such an extent that the pollen cell loses its character and becomes callus. Therefore, in short, plant growth regulators need to be avoided in
pollen and anther culture medium.

Modification of sugar and amino acids is more important in embryogenesis than plant growth regulators. Nitrogen metabolism is
quite an important feature of the in vitro cultures, especially when the
technique is used for developmental purposes. The presence of ni-
trate or of ammonium salt in the medium cannot, in some instances,
replace the need for amino acids, which appear to play very specific
roles at different stages of the developmental process. When the technique is used to achieve the complete cycle of development from a
single cell to a complete plant, it is important not to overlook these
substances. Here again, no general rule can be given, and empirical
studies have to be carried out with each plant species.

The environmental conditions under which the cultures are to be
placed can enhance the differentiation of the globular embryos into
plantlets with cotyledons and roots (Dunwell, 1976).

1. Liquid medium seems to be more favorable than solid medium,
as this favors further development of the embryo.
2. Another important environmental factor is light. While darkness and blue light were found to be promotive, white light was
inhibitory in androgenesis. Red light shortens the time necessary to produce plants from pollen by 20 percent.
3. Activated charcoal plays an important role in androgenesis; it
absorbs not only toxic substances from degenerated anthers, but
also plant growth inhibitors such as abscisic acid (ABA) which

are present at high concentrations in anther and are known to prevent embryogenesis.

## Importance of Androgenesis

In a plant breeding program of selection, the time necessary to have an isogenic (genetically identical) line is reduced from a minimum of five or six generations to one or two using the technique of androgenesis. If the doubling of the chromosome is achieved by treating the anther with colchicine at the time of the first pollen mitosis just before the in vitro culture, the homozygous line is obtained from the first generation. When the culture is done with the untreated microspore, the plant produced is haploid and therefore sterile and its chromosome number needs to be doubled. Therefore, only two generations of plants are needed to produce an isogenic line, which can be useful for traditional plant breeding purposes.

Homozygotes induced in tissue culture following meiotic segregation could reveal a number of valuable recessive characters which have accumulated and remained unexpressed in natural heterozygous populations. Many valuable genotypes, such as those governing low nutritive requirements and resistance to cold, drought, or heavy metals, are of this type, and their expression in spontaneous or induced mutants will be promptly detected.

In trees, widespread self-incompatibility can be overcome through androgenesis. Furthermore, in the field of cell fusion and regeneration, protoplasts of haploid cells are ideal material for gene transfer.

## Regeneration of Diploid Plants

Regeneration of diploid plants can either occur by spontaneous doubling of chromosomes by endomitosis (multiplication of chromosomes without nuclear division); or by treatment with colchicine. Colchicine is an alkaloid isolated from autumn crocus, *Colchicum autumnale,* family Liliaceae, and is a powerful poison. Extreme care should be exercised when handling this chemical. Colchicine specifically binds tubulin (a kind of protein) molecules. The inactivation of this structural protein blocks microtubule polymerization and the formation of the spindle apparatus during the metaphase leading to the failure of separation of the sister chromatids. This results in a doubling of the number of chromosomes in the nucleus of the cells at the end of mitosis. To induce diploidy when plantlets are beginning to emerge from the cultured anthers, the anthers are immersed in an aqueous solution of colchicine (0.5 percent w/v) for 24 to 48 hours (Figure 4.3). Following the colchicine treatment, the anthers containing the young plantlets should be rinsed in double distilled water and be cultured on agar medium. It is important to ascertain the chromosome numbers of the plantlets arising from such anthers with acetocarmine or Feulgen staining. If diploidy is not achieved, the immersion time may be increased to 96 hours, or the treatment may be repeated to achieve success.

## Pollen Dimorphism

It is important to note that the effective number of embryogenic or callogenic pollen is notoriously low. By far the majority of pollen follow the gametophytic mode of development, which is indeed their

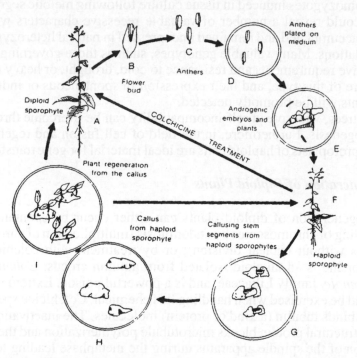

FIGURE 4.3. Methods of obtaining homozygous diploids from androgenic haploids. (*Source:* Bhojwani and Razdan, 1983, p. 138.)

natural destiny, while the fraction of pollen giving a sporophytic response appears to be intrinsically limited. Populations of pollen could therefore contain a predestined atypical class, and much effort has been directed toward finding a means of recognizing this minority in the herbaceous species that are commonly used for studies of androgenesis.

Sunderland and Wicks (1971) recorded atypical pollen grains lacking starch and poor in cytoplasm. It was then postulated that these were significant for the induction of sporophytic growth. Later, a correlation between the frequency of pollen callus or embryos, their distribution within and among anthers, and apparent similarities and development in situ and in vitro was established. Centrifugal separation of two classes of pollen and subsequent culture ab initio later confirmed the validity of the present notion of pollen dimorphism and its relevance to androgenesis.

Work in woody plants also confirms the notion of pollen dimorphism. For example, Thanh-Tuyen and de Guzman (1983) noted that embryogenic microspores in coconuts were smooth, fully expanding, and had a lightly staining cytoplasm, whereas darkly stained grains with deposits were considered to be nonembryogenic.

The observation of pollen dimorphism points to the existence of a class which is immediately programmed for androgenesis, but, however predetermined this propensity may be, future investigations need to interfere with, and regulate, the process, thus improving androgenic potential of pollen grain.

## Albinism

The occurrence of albinos among embryos produced in vitro, whether of somatic or androgenic origin, has been repeatedly reported in the literature. This phenomenon, which is due to impaired chlorophyll synthesis, was first reported by Chinese workers in rice. In general, this undesirable characteristic, which causes problems in breeding programs, is very common in Poaceae, but the occurrence of albino embryos is rare in Solanaceae. Examples of albino embryoids in vitro have also been reported in woody plants such as *Gingko biloba* and *Populus* sp.

# Chapter 5

# In Vitro Pollination and Fertilization

## *DEVELOPMENT OF A FEMALE GAMETOPHYTE*

In angiosperms a limited number of cells are produced in the female gametophyte (embryo sac). Although the range in the number of nuclei in the mature gametophyte is between 4 and 16, close to 80 percent of angiosperms have 8 nuclei.

Several types of female gametophytes are recognized on the basis of:

1. The number of spores that take part in the development
2. The number of mitotic divisions after the two meiotic divisions
3. The total number of nuclei at the completion of division
4. The cellular organization of the mature gametophyte

The most common type is eight-nucleate (polygonum) type (Figure 5.1):

1. The egg is the most important, conspicuous, and relatively uniform structure. It is present at the micropylar end in close association with the synergids.
2. The synergids seem to play an active role in nutrition of the embryo sac, attracting the pollen tube, and acting as an osmotic buffer for the release of pollen tube contents.
3. The binucleate central cell or polar nuclei is the storehouse of starch, sugars, amino acids, and inorganic salts. The central cell acts as an endosperm mother cell after fertilization.
4. Three antipodal cells appear to be metabolically very active, implying a nutritive role.

57

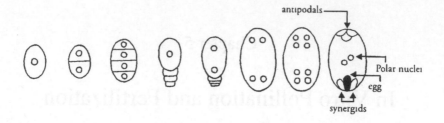

FIGURE 5.1. Diagram illustrating polygonum embryo sac development.

## POLLINATION

Pollination is commonly defined as the process of pollen transfer from anther to stigma of the same flower or another flower of the same species.

If the pollen is compatible, it germinates on the stigma, producing a pollen tube that carries the male gametes to the female gametophyte, consequently resulting in the union between male and female gametes. A certain group of pollen-borne proteins called lectins, located in both exine and intine, seems to play an important role in the pollen-pistil recognition mechanism.

If the pollen is incompatible, then either its germination is prevented or the pollen tube growth is arrested in the transmitting tissues. This incompatibility can be manifested in the tissues of either the stigma or the style at various stages before fertilization.

Because it is often necessary in plant breeding work to overcome incompatibility, several methods have been developed to this end with varying degrees of success. These methods include:

- Bud pollination
- Delayed pollination
- In vitro pollination
- Intra ovarian pollination

Longevity of the pollen is constant at species level. When pollen grains are ready for dispersal, they are in a state of dormancy with a water content of 10 to 15 percent—similar to that of seeds. Pollen grains of grasses are short-lived. For example, *Paspalpum* pollen loses its viability within 30 minutes. In the majority of angiosperms, pollen viability drops dramatically 24 hours after dehiscence. However, pollen life can be artificially prolonged by storage in low temperature and low relative humidity.

Pollen germination and growth in most angiosperms can occur immediately or within a few minutes of release from the anther, provided there is an adequate supply of water, presence of certain inorganic salts including boron, and a source of energy such as sucrose.

In a typical angiosperm, when the pollen lands on the stigmatic surface it adheres to the stigmatic exudate and then germinates to produce the pollen tube. Initially, the tube is only an extension of the intine, but it soon acquires terminal growth. During its growth, a series of callose plugs (a polysaccharide particularly found in the sieve plates of sieve tubes) isolate the protoplasm of the growing tip from the empty tube behind.

In most angiosperms in which pollen is shed at the two-celled (binucleate) stage, division of the generative cell into two male gametes occurs in the growing tube. The tube penetrates the stigma, grows in between the cells of the transmitting tissue in the style, and reaches the ovule.

The interval between pollination and fertilization is usually between 12 and 24 hours. In *Taraxacum* (dandelion) it can be as short as 15 minutes, whereas it could be several weeks in some orchids (in which ovules start differentiating only after pollination) and even 12 to 14 months in *Quercus* (oak).

## FERTILIZATION

In angiosperms, both male gametes brought in by the pollen tube are involved in fertilization. One fuses with the egg to produce the embryo and the other fuses with the polar nuclei to produce the endosperm. This process is termed double fertilization (Figure 5.2).

Sexual hybridization is a powerful tool for producing superior plants by combining characteristics distributed in different individuals (or even species) of a genus. The technique involves controlled, artificial pollination of the female parent with pollen from the selected male parent.

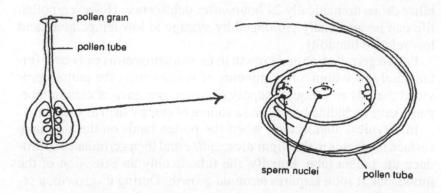

FIGURE 5.2. Double fertilization in an angiosperm.

Transferring viable pollen from one parent to the receptive stigma of another does not always lead to seed setting. There are two kinds of barriers to fertilization:

1. Prefertilization or prezygotic:
   • inability of pollen to germinate on stigma
   • failure of the pollen tube to reach the ovule due to excessive length of the style, or slow growth of the pollen tube which fails to reach the base of the style before the ovary abscises
   • bursting of the pollen tube in the style.
2. Postfertilization: fertilization may occur normally, but the hybrid embryo fails to attain maturity due to embryo-endosperm incompatibility or poor development of the endosperm.

## In Vitro Pollination, Seed Set, and Embryo Rescue

Pollination followed by fertilization normally leads to the production of an embryo which, in the intact plant, is linked with normal seed development. Most angiosperms are out-breeders, thus self-pollination is limited. Furthermore, interspecific (between species) and intergeneric (between genera) hybridizations are also rare in nature.

In plant breeding, however, selfing and hybridization are methods commonly used to obtain desirable crosses. In vitro pollination and fertilization is a powerful experimental approach which aids in achieving crosses. Since many of the embryos obtained in this manner will not survive, they must often be rescued. This chapter will discuss these two aspects of in vitro culture.

## *Terminologies*

- In vitro ovular pollination is application of pollen to excised ovules.
- In vitro placental pollination (Figure 5.3) is application of pollen to ovules attached to the placenta.
- In vitro stigmatic pollination (Figure 5.4) is application of pollen to the stigma.

The technique of in vitro pollination and fertilization was pioneered by Indian scientists Kanta, Rangaswamy, and Maheshwari in 1962 on opium plant *(Papaver somniferum)*. Since that time, several species have been pollinated successfully and fertilized in vitro including *Swainsona laxa,* a rare Australian plant (Taji and Williams, 1987).

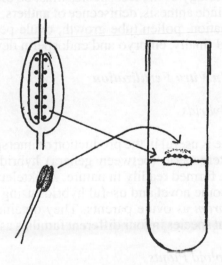

FIGURE 5.3. Schematic illustration of placental pollination in vitro. (*Source:* Pierik, 1987, p. 240, fig. 22.1, with kind permission from Kluwer Academic Publishers.)

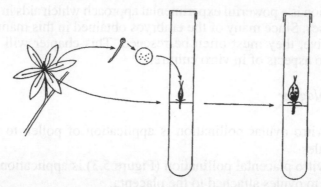

FIGURE 5.4. Schematic illustration of stigmatic pollination in vitro. (*Source:* Pierik, 1987, p. 241, fig. 22.2, with kind permission from Kluwer Academic Publishers.)

Success of this technique depends on two basic considerations:

1. Using pollen grain and ovules at the proper developmental stage.
2. Obtaining nutrient media that will support pollen germination, pollen tube growth and embryo development. Thus, many aspects of the floral biology of the test species must be understood. These aspects include anthesis, dehiscence of anthers, pollination, pollen germination, pollen tube growth, ovule penetration, fertilization, and finally, embryo and endosperm development.

### Application of In Vitro Fertilization

#### Production of Hybrids

This technique is useful in the production of interspecific (between species) and intergeneric (between genera) hybrids, which would otherwise not be formed readily in nature. Zenkteler (1970) successfully produced some novel and useful hybrids using *Melandrium album* and *M. rubrum* as ovule parents. They obtained hybridization using 15 different species in four different families as pollen parents.

#### Induction of Haploid Plants

As discussed in anther culture, techniques such as delayed pollination, distant hybridization, pollination with abortive or irradiated pollen, and physical and chemical treatments to the ovary have been used to induce haploidy. More recently, anther (or microspore) culture has been tried with more success. It is possible to obtain haploid plants through in vitro pollination and fertilization. This has been achieved in *Mimulus luteus* when the ovule mass was pollinated with pollen from *Torenia fournieri* (both from the same family, Scrophulariaceae). One percent of plants produced were haploid.

## Overcoming Sexual Self-Incompatibility

Self-incompatibility is a physiological barrier which prevents fusion of sexually-different gametes, which are otherwise fertile, produced by the same individual of a heterosporous species.

In this process, the entire ovule mass of an ovary, intact on the placenta, is cultured after pollination. This allows the original in vivo arrangement of the ovules to be retained.

## Pollen Physiology and Fertilization

These techniques of in vitro pollination and fertilization can be used in the studies of pollen physiology and fertilization. Using these techniques, Balatková and Tupy (1968) were able to determine that the pollen tube of *Nicotiana tabacum* affects fertilization even after gamete formation in pollen culture.

The measure of success of in vitro fertilization and subsequent production of viable seeds is dependent upon the following factors:

1. Age of the explants, particularly that of the ovule
2. Adequate pollen germination
3. Proper growth of pollen tubes and gametogenesis (development of sperm)
4. Pollen tube entry into ovules
5. The degree of fertilization

There is a paucity of information regarding the effect of environmental conditions, such as light, temperature, and humidity on in vitro pollination. Furthermore, factors such as the location of placement of the pollen, and the ability to keep the ovule and placenta surface free of water may play an important role in achieving success in in vitro pollination. In addition to these factors, success is also influenced by factors involved in the development of the zygote into mature viable embryos and of the fertilized ovule into seeds.

The constituent of the nutrient medium is obviously very important. One component that has had a beneficial effect in some cases has been the addition of casein hydrolysate (500 milligrams per liter). Casein hydrolysate is an undefined mixture of amino acids occasionally used in nutrient media. The milk protein, casein, is hydrolyzed (treated with an enzyme or acid) to form a complex mixture of amino acids.

It is clearly possible for normal development of viable seeds to occur. However, in some cases, embryo degeneration and other abnormal events take place. Under these conditions, the technique of embryo rescue (i.e., excising the immature embryo and allowing it to grow and develop into a viable mature embryo on culture medium) can be applied.

## EMBRYO CULTURE

The underlying principle of embryo culture is the aseptic excision of the embryo and its transfer to a suitable medium for development under optimum culture conditions (Figure 5.5). In general, it is relatively easy to obtain axenic embryos, since they are within the sterile environment of the ovule. Thus, entire ovules or immature seeds are sterilized and the embryos are aseptically separated from the surrounding tissue. Seeds with hard seedcoats are generally surface-sterilized and then soaked in water for a few hours to a few days. In the latter case, they are surface-sterilized again before embryo excision. Splitting the seeds under aseptic conditions and transferring the embryo directly to a nutrient agar medium is the simplest technique used. With smaller embryos it is important that they be removed uninjured. This often requires the use of a dissecting microscope. In the case of orchids, the entire ovule is cultured since there are no functional storage tissues and the seedcoat is reduced to a membranous structure. It is important to keep the excised embryo moist at all times to avoid desiccation during the operation. The most critical aspect of embryo culture work is the selection of the medium necessary to sustain continued growth of the embryo. The media formulations used over the years vary tremendously and many of them have not been rigorously determined. Certain generalizations, however, can be made.

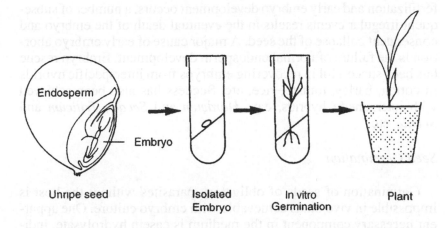

FIGURE 5.5. Schematic representation of zygotic embryo culture.

The younger the embryo, the more complex is its nutrient requirement. Relatively mature embryos can be grown in a much simpler inorganic salt medium supplemented with a carbon (energy) source such as sucrose. Relatively young embryos require, in addition, different combinations of vitamins, amino acids, growth hormones and, in some cases, natural endosperm extracts such as coconut water. Coconut water (the liquid endosperm of coconut) is an undefined medium used with success in early plant tissue culture and it is still being used. Its use was first reported by Van Overbeek in 1942 in tissue culture of *Datura* embryo. Coconut water is known to contain plant growth regulators, especially zeatin and myoinositol. Myoinositol, one of the components of vitamin B complex, is part of various membranes, particularly those of the chloroplast, and is also involved in the synthesis of phospholipid and cell wall pectins. Since embryos are often embedded in the ovular sap under considerable osmotic pressure, the presence of an osmoticum, such as mannitol, in culture medium is also recommended.

## Applications of Embryo Culture

### Overcoming Embryo Inviability

Embryo abortion occurs quite frequently as a result of unsuccessful crosses in breeding. Although in most of these cases successful fertilization and early embryo development occurs, a number of subsequent irregular events results in the eventual death of the embryo and consequent collapse of the seed. A major cause of early embryo abortion is the failure of normal endosperm development. Embryo rescue has been successful in recovering embryos from interspecific hybrids in cotton, barley, tomato, rice, etc. Success has also been achieved with intergeneric hybrids, e.g., *Hordeum* and *Secale, Triticum* and *Secale,* etc.

## Seed Germination

Germination of seeds of obligatory parasites without the host is impossible in vivo but is achievable with embryo culture. One apparent necessary component in the medium is casein hydrolysate, indicating specific amino acids may well be required from the host for germination.

## Overcoming Seed Dormancy

The causes of seed dormancy are varied and include the presence of endogenous inhibitors, specific light, temperature, storage requirements, and the degree of maturity of the embryo. Embryo culture has been successfully used to overcome dormancy due to inhibitors in *Iris* seeds, and for successful germination of immature orchid seeds. Seed dormancy may be imposed by seedcoat and/or endosperm. By removing these, the seeds germinate immediately. Slow seed germination, due to restriction by seedcoat for $O_2$ and $H_2O$ uptake, can also be eliminated by embryo culture. Embryo culture, therefore, speeds up germination. One must bear in mind that culture of immature embryos can result in an even further shortening of the breeding cycle.

## Monoploid Production

Production of monoploids (haploids) occurs in cereal, and specifically in barley with the cross between *Hordeum vulgare* and *H. bulbosum.* Fertilization occurs, but thereafter the chromosomes of *H. bulbosum* are eliminated. The result is that the haploid embryo of *H. vulgare,* which is only viable by embryo culture, remains. After chromosome doubling, a homozygote *H. vulgare* is finally obtained.

## Prevention of Embryo Abortion
with Early Ripening Stone Fruits

In crosses of early ripening stone fruits (e.g., peach, cherry, apricot, plum) the transport of water and nutrients to the immature embryo is sometimes cut off too soon, resulting in abortion of the embryo. In practice, this means that no crossing with an early ripening parent is possible. Therefore, embryo rescue is the only option to overcome this problem.

## Vegetative Propagation

In Poaceae and conifers, embryos are often used as the starting material for vegetative propagation. They appear to be very responsive to manipulation because they are juvenile. With plants from Poaceae, organogenesis takes place relatively easily from callus arising from juvenile tissue while, with conifers, the cloning via immature callus derived from young embryo and axillary shoot formation is also easy.

## Experimental Embryogenesis

In addition to the applied use of embryo culture, experimental embryogenesis has proven useful in the study of the growth requirements of embryos, particularly the role of phytohormones, nutrients, and environmental conditions on zygotic embryogenesis.

## Problems of Embryo Culture

One major problem in embryo culture is precocious germination. In this phenomenon, excised immature embryos of some species do not continue normal embryo development in culture but grow directly into small weak plantlets. It appears that an interplay of inhibitors, high osmotic pressure, and possibly low $O_2$ tension in the ovule may be involved in the in vivo regulation of normal embryo development.

The technique of in vitro pollination appears very promising for overcoming fertilization barriers to compatibility and creating new genotypes. Although the feasibility of the technique was demonstrated over 30 years ago, not much interest has been shown by plant breeders in applying it to a specific desirable cross. More novel areas of plant tissue culture such as protoplast fusion technology and genetic engineering have probably overshadowed this important area of research.

# Chapter 6

# Somatic Hybridization
# Using Protoplast Technology

## INTRODUCTION

One of the most striking characteristics of plant cells is the presence of a thick and relatively rigid cell wall. The cell wall provides mechanical support and is used in defense against physical damages and attack by pathogens. It may also play a role in cellular communication.

One of the problems caused by the presence of a cell wall is that it prevents direct access to the cell membrane and, therefore, impedes certain manipulations. For example, the study of membrane function through the use of molecular probes such as antibodies is not possible directly on plant cells.

Biotechnological manipulations, such as cell-cell fusion, are also precluded due to the presence of the cell wall. An additional problem caused by the presence of a cell wall is that single cells, as opposed to multicellular tissue fragments or cultured cell clusters, are difficult to obtain. This is because the wall components and intercellular polymers effectively cement cells together. Because of this, single cell cloning or selection of desirable cell types in plants can be difficult.

In 1960, E.C. Cocking made the crucial discovery that cellulose-degrading enzymes were able to dissolve away the cell wall and produce the fragile but still viable protoplast. Of course, protoplasts had been isolated before that, for example, by chopping up plant materials causing some intact protoplasts to escape through the ruptured walls. But, the great power of the enzymic isolation technique was that it allowed, for the first time, the isolation of protoplasts in numbers large enough to be suitable for further experimentation.

Following the first isolation of protoplasts, advances in the enzymic isolation technique came quickly. In the 1960s, using the proper nutritional requirement, protoplasts were induced to divide and this line of research culminated in the report of the regeneration of whole plants from the protoplast in 1972 (Carlson, Smith, and Dearing, 1972). This result was of great fundamental importance to biotechnology for it was now shown that plant species with "improved and desired" characteristics could, in principle, be recovered from single protoplasts, following the appropriate manipulation or selection.

The protoplast, therefore, appeared to offer the ideal experimental material for a range of studies in plant cell biology and biotechnology. However, as more became known about the structure and properties of protoplasts, the initial enthusiasm diminished, tempered by a recognition of the difficulties encountered in their use. Furthermore, doubts were raised concerning the suitability of protoplasts as models for the intact plant cell, an important consideration for fundamental research into plant biology.

However, over 30 years of protoplast research has led us to the position where a realistic view of the potential and limitations of protoplasts is possible. Some notable achievements have already been made, some areas of early promise have remained unfulfilled, but most researchers agree that protoplasts play an important role in the investigation of the biology of plant cells and in their use in biotechnological research and in vitro breeding.

## Definition

Isolated protoplasts have been described as "naked" plant cells because the cell wall has been experimentally removed. This distinction is important to make because we cannot assume that the removal process does not affect the properties of the protoplast.

# USES OF PROTOPLAST TECHNOLOGY

1. Two or more protoplasts can be induced to fuse and the fusion product carefully nurtured to produce a hybrid plant. Although this phenomenon has been observed repeatedly, fusion has not been achieved with the isolated protoplasts of some species. In some cases, hybrids that cannot be produced by conventional

plant genetics because of sexual or physical incompatibility can be formed by somatic cell fusion.

2. After removal of the cell wall, the isolated protoplast is capable of ingesting "foreign" material into the cytoplasm by a process similar to endocytosis (e.g., as in amoeba). Experimentally, some progress has been made on the introduction of nuclei, chloroplasts, mitochondria, DNA, plasmids, bacteria, viruses, etc.

3. The cultured protoplast rapidly regenerates a new cell wall and this developmental process offers a novel system for the study of cell wall biosynthesis and deposition.

## *OBTAINING PROTOPLASTS*

The chief function of the cell wall is to exert a pressure on the enclosed protoplast and thus prevent excessive water uptake leading to bursting of the cell. Before the cell wall is removed, the cell must be bathed in an isotonic solution, which is carefully regulated in relation to the osmotic potential of the cell. In general, mannitol or sorbitol (13 percent w/v) has given satisfactory results. These sugar alcohols are either not metabolized or slowly metabolized. Sugars, such as sucrose, have not been used because as they are taken up by the cells their effective concentration falls. This situation may cause problems with protoplast stabilization but it has also been exploited to provide the progressively declining osmoticum value that many protoplast types require as they undergo division and colony formation. Salt solutions have also occasionally been used as osmotica.

In general, in developing a technique, a range of concentrations of mannitol from 5 to 15 percent (w/v) may need to be tested. It is important to note that a preparation bathed in a plasmolyticum of too low a concentration may lead to multi-nucleate protoplasts, owing to the spontaneous fusion of two or more protoplasts during the isolation procedure.

Protoplasts can be released from the cells by either a mechanical or an enzymatic process. The mechanical approach involves cutting a plasmolyzed tissue in which the protoplasts have shrunk and pulled away from the cell wall. Subsequent deplasmolysis results in expansion and release of the protoplasts from the cut ends of the cells. In practice, this technique is very difficult and yields only a few viable protoplasts per exercise. One advantage, however, is that the complex

and often deleterious effects of the wall-degrading enzymes on the
metabolism of the protoplasts are eliminated.

Since the isolation of cellulose-degrading enzymes from wood-
rotting fungi in 1960 (Cocking), almost all protoplast isolation work
has been performed enzymatically. By using enzymes, a high yield
($2\text{-}5 \times 10^6$ protoplast/g leaf tissues) of protoplasts is obtained after removal of cellular debris. The technique consists of:

1. Surface sterilization of leaf tissues
2. Incubating the tissue in a suitable osmoticum
3. Peeling off the lower epidermis or slicing tissue to aid enzyme
   penetration
4. Sequential or mixed-enzyme treatment (Figure 6.1)
5. Purification of the isolated protoplast by removal of enzymes
   and cellular debris
6. Transfer of the protoplasts to a suitable medium with the appropriate culture conditions (Figure 6.2)

The plant cell wall consists of a complex mixture of cellulose,
hemicellulose, pectin, and lesser amounts of protein and lipids. Because of the chemical bonding of these diverse constituents, a mixture of enzymes would appear necessary to effectively degrade the
system. Cellulose is a polymer of D-glucose. The big bulk of hemicellulose fraction in angiosperms are xylans. These polymers consist
of several monosaccharides in addition to xylose. Pectins are poly-
saccharides containing the sugars galactose, arabinose, and the galactose derivative galacturonic acid. Accordingly, most protoplast isolation procedures include a cellulase, hemicellulase, and pectinase, or a
crude enzyme preparation containing all three (Table 6.1). Proto-
plasting enzymes are isolated from a number of sources, usually fungi
such as *Aspergillus niger* or *Trichoderma viride*. Currently, a large number of enzymes are available commercially, all showing slightly different
substrate specifications, from exotic sources such as snail gut.

Much of the uncertainty in protoplast isolation may be due to sub-
tle differences in wall structure between species and between plants
or cells grown under different conditions. Unfortunately, the elucida-
tion of precise wall configuration is an extremely difficult biochemical problem.

Protoplasting-enzymes are rarely pure. They may contain ribo-
nucleases, proteases, and several other toxic enzymes. Because of the

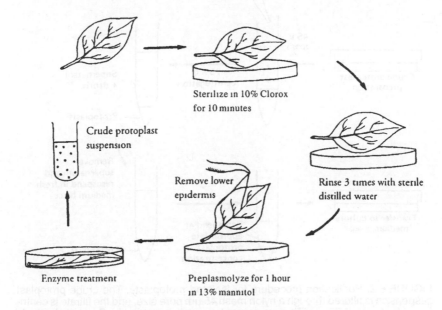

Sterilize in 10% Clorox
for 10 minutes

Crude protoplast
suspension

Remove lower
epidermis

Rinse 3 times with sterile
distilled water

Enzyme treatment

Preplasmolyze for 1 hour
in 13% mannitol

FIGURE 6.1. Basic techniques for the isolation of protoplasts from an excised leaf. The leaf is surface sterilized, rinsed repeatedly in sterile distilled water, and the cells are plasmolyzed in a solution of mannitol. The lower epidermal layer is stripped from the leaf to enhance enzyme penetration into the mesophyll tissue. Following treatment with one or more wall-degrading enzymes, a crude suspension of mesophyll protoplasts is obtained. (*Source:* Dodds and Roberts, 1985, p. 136 Reprinted with the permission of Cambridge University Press.)

presence of these deleterious enzymes, along with other harmful substances, several purification procedures have been developed (e.g., gel filtration), but in many cases, especially with the more amenable species, this is not usually necessary. It is important to minimize the contact time of the plant material with the enzyme solution to reduce the damage caused by unwanted contaminants. Incubation times vary from one hour to overnight. Typically, a four-hour incubation is used for most species.

## Pretreatment of Plant Material

The condition of the material from which protoplasts are to be isolated is critical to the success of the procedure. The factors that have

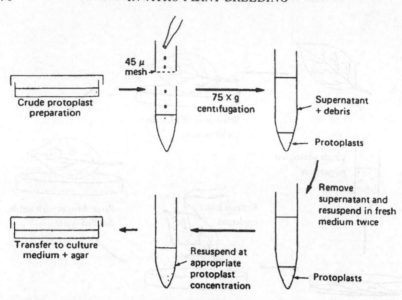

FIGURE 6.2. Purification procedure for isolated protoplasts. The crude protoplast suspension is filtered through a nylon mesh 45 μm pore size, and the filtrate is centrifuged for 5 min at 75 × g. The supernatant, carefully removed by Pasteur pipette, is discarded. The protoplasts, resuspended in 10 ml of fresh culture medium, are again centrifuged. Once again the supernatant is removed. The centrifugation and resuspension process is done three times. Before transfer of the protoplasts to a culture medium, the preparation is examined for protoplast density and viability. (*Source:* Dodds and Roberts, 1985, p. 137. Reprinted with the permission of Cambridge University Press.)

TABLE 6.1. Some Enzyme Preparations Exhibiting Wall-Degrading Activity, Classified According to Major Function

| Cellulases | Hemicellulases | Pectinases |
|---|---|---|
| Cellulase Onozuka (R-10, RS) | Rhozyme HP-150 | Macerase Macerozyme R-10 |
| Cellulysin | Hemicellulase (Sigma) | Pectolyase Y-23 |
| Meicelase (CESB, CMB) | | Pectinase (Sigma) |
| Driselase | | |

so far been recognized as important in this respect include temperature, lighting, humidity, and tissue age.

At present, the selection of the appropriate growth conditions for tissue used in protoplast isolation is largely a matter of trial and error, because the specific cellular variables involved are poorly understood.

The age of the plant is usually important to the success of protoplast isolation. It has been reported for tobacco plants that leaves must be taken from plants 40 to 60 days after germination. These leaves produce optimum protoplast yield and high survival rate.

After protoplasts are obtained, they have to be purified. This is carried out by filtration (through stainless steel or nylon sieves, 0.45 μm) to remove impurities such as cell walls and remains of the cells, etc., followed by repeated centrifugation in a sugar solution to pellet the protoplasts. Because protoplasts are extremely delicate, they have to be centrifuged at a low speed under which conditions the intact protoplasts float on a 15 to 20 percent sugar solution, while the debris will form pellets at the bottom of the centrifuge tube.

## A Typical Protoplast Isolation Protocol

The following is the outline of a method used to prepare protoplasts from tobacco leaves:

- Select young, fully expanded leaves from tobacco leaves.
- Cut approximately 0.5 g leaf tissue (avoiding major veins, which do not yield protoplasts) into 2 mm strips using a scalpel.
- Float the strips on an enzyme solution containing 1 percent (w/v) cellulase, 0.5 percent hemicellulase, 0.5 percent pectinase, and 0.4 M sorbitol buffered at pH 7.5.
- Incubate statically overnight at 25°C.
- Gently shake the suspension to disperse digested tissue.
- Filter through 40 to 50 μm nylon mesh.
- Centrifuge filtrate at 100 g for five minutes.
- Disperse pellet in 3 mL 23 percent sucrose solution and centrifuge for five minutes at 100 g.
- Remove floating layer (containing viable protoplast) and wash three times (by centrifugation) in 0.4 M sorbitol solution at pH 7.5.

## Determination of Protoplast Density and Viability

Before the isolated protoplasts can be placed into culture, it is necessary to examine them for viability with fluorescein diacetate. This dye, which accumulates only inside the plasmalemma of viable protoplasts, can be detected by fluorescence microscopy. Another stain that can be used to test protoplast viability is Evans blue. Intact viable protoplasts, by virtue of their intact plasmalemmas, are capable of excluding this biological stain. Impermeability of the cell to this dye presumably indicates a living cell. In addition, cyclosis or protoplasmic streaming has been reported to be a measure of viability.

Protoplasts have a minimum and maximum plating density for growth. The optimum plating efficiency for tobacco protoplasts is about $5 \times 10^4$ protoplasts per cubic centimeter. The protoplasts fail to divide when plating density is one tenth of this concentration. Other protoplasts have a similar range, e.g., petunia's optimum plating density is $2.5 \times 10^4$.

The concentration of protoplasts in a given sample can be determined by the use of a Fuchs-Rosenthal modified hemacytometer with a field depth of 0.2 mm. It is important to note that haemocytometers designed for blood counts, are not suitable because the field depth of 0.1 mm is too shallow for protoplasts isolated from large plant cells. By use of this instrument, it is possible to adjust the concentration of protoplast to the appropriate level. Because the protoplast preparation will be diluted by an equal quantity of agar-containing medium, the sample should be adjusted to a concentration of $10^5$ protoplasts per cubic centimetre.

## Protoplast Fusion

Isolated protoplast membranes are negatively charged relative to the surrounding medium. This charge causes electrostatic repulsion between adjacent protoplasts and between protoplasts and other negatively-charged particles, for example, certain viruses and DNA. This electrostatic force of repulsion must be overcome before protoplasts can be fused together to produce useful hybrids, and before the protoplasts can be induced to take up negatively-charged macromolecules.

To obtain efficient fusion, it is necessary to work with a large number of viable protoplasts. Fusion is generally induced by incubation in a high concentration of polyethylene glycol (PEG), a high concentration of $Ca^{2+}$, and relatively high pH (5.5 to 6.5). This treatment is necessary to remove the negative charge on the protoplasts. After fusion has taken place, PEG and $Ca^{2+}$ ions should be washed away.

In recent years, electrofusion (electrically induced fusion) has become popular. This technique yields more viable protoplasts and hence better fusion than PEG. Electrically induced fusion is directed by exposing protoplasts to alternating electrical fields. This results in protoplast membranes becoming differentially charged causing positive regions to be attracted to negative regions on adjacent protoplasts.

If different protoplasts fuse (originating from two different genotypes or plant species), then a heterokaryon results, and if two similar protoplasts fuse, then a homokaryon is obtained. After cell fusion has taken place, nuclear fusion can occur when the nuclei divide simultaneously.

After fusion has taken place, the biggest problem is to select somatic hybrids from the cell population. If, for example, hybridization is carried out between protoplasts of two sorts A and B, then a population will be obtained which contains nonfused A and B protoplasts, AA and BB homokaryons, AB heterokaryons and diverse multiple karyons (random fusion of several protoplasts). Many selection procedures have been developed to efficiently select only the AB heterokaryons from this mixture, e.g., use of a fluorescence-activated cell sorter (FACS, Figure 6.3). In this case, the different parental protoplast types are labelled with different fluorochromes, commonly fluorescein and rhodamine, which fluoresce at different wavelengths.

In the FACS, a given fluorescence activates a switch that directs the flow of liquid through the machine to one of a number of reservoirs. The dual fluorescence present in fusion products can thus be used to isolate them. One problem with this technique arises because adhering (not fused) protoplasts and multiple fusion products may all be recognized by the machine. Also, passage through the machine may cause extensive cellular damage due to the fragility of protoplasts. This problem can be minimized if the protoplasts are first cultured for 24 to 48 hours after fusion so that the regenerating wall affords some protection against physical damage in the FACS.

## THE CULTURE OF PROTOPLASTS

The culture of protoplasts has very specific requirements because the ensuing development consists of many different stages:

- Regeneration of the cell wall
- Cell division
- Callus formation
- Organ and/or embryo differentiation

During regeneration of the cell wall, the $Ca^{2+}$ concentration should be high, while the osmotic potential of the nutrient medium should be low. The protoplast density (as is the case for cell suspension culture) in the culture medium also plays an important role in cell wall regen-

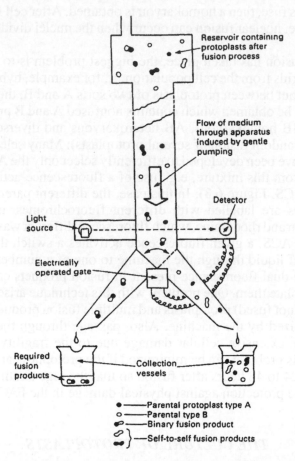

FIGURE 6.3. A simplified diagram of a fluorescence-activated cell sorter (FACS). (*Source:* Stafford and Warren, 1991, p. 71. © John Wiley & Sons Limited. Reproduced with permission.)

eration. After regeneration of the cell wall, which happens within the first few days, the osmotic potential of the medium must be gradually increased. The first cell divisions usually take place two to seven days after protoplast isolation.

The culture of protoplasts can take place in different ways. In solid media, agarose is usually used as the gelling agent. Sea plaque agarose of marine colloids is strongly recommended as it is much less toxic than agar. Protoplasts are cultured by any of the following methods:

1. Suspended in a thin layer of liquid medium in a Petri dish
2. On solid medium
3. Grown in microchambers between glass plates
4. In drops of liquid medium (microdrop method) which are placed in a Petri dish
5. In "nurse culture," especially when the density of protoplast is low. The feeder layer technique may be used instead of nurse culture, hence the protoplasts are grown on a cell layer from which they are separated by cellophane or filter paper
6. In a layer of liquid medium on top of a solid layer (bilayer technique)

## Factors Affecting Successful Regenerating from Protoplasts

1. The family, the species, the genotype (members of Solanaceae are by far the most suitable group of plants for protoplast culture).
2. The state of differentiation of the donor cells; juvenile cells and cells from shoot tips have greater regenerative potential.
3. The growth conditions of the starting material.
4. The choice of medium, e.g., richer media for formation of callus, weaker media for formation of embryoids.

## Advantages of Protoplast Fusion and Somatic Hybridization

Using somatic hybridization it is possible to create hybrids that would not be recoverable by normal crossings (due to taxonomic or sexual barriers). An interesting example is the somatic hybridization between *Solanum tuberosum* (tuber forming plant with little or no disease resistance) and *Solanum brevidens* (nontuber forming plant but disease resistant). Due to sexual incompatibilities it is impossible for such hybridization to occur in vivo. However, it is important to note that while theoretically we can produce somatic hybrids through protoplast fusion from some completely unrelated and incompatible species, this form of hybridization should be avoided because the resultant produce is usually unstable. Somatic hybridization can be used as an alternative to colchicine treatment for obtaining tetraploids.

## Genetic Consequences of Protoplast Fusion

The most obvious result of protoplast fusion is an additive increase in the chromosome complement of the hybrid cell. This situation rarely, if ever, persists when the protoplast is cultured, and when a cell colony or eventually a plant is derived, it is usually found that some or most chromosomes from one parent are preferentially lost.

The mechanisms governing which set of chromosomes is to be partially eliminated are not well understood. However, the concept of genome competition seems to play an important role. As a general rule, the chromosomes of the parent with the shorter cell cycle time are often retained. This suggests selection on the basis of direct competition. Somatic hybrids that lose chromosomes from one parent and, therefore, are phenotypically closer to the other parent, are called asymmetric. The rare hybrids that retain substantial amounts of genetic material from both parents are termed symmetric.

The most widely known somatic hybrid is 'Arabidobrassica' a cross between *Arabidopsis thaliana* (10 chromosomes) and *Brassica campestris* (20 chromosomes), both of the Brassicaceae family. The hybridization of these species resulted in the formation of a range of products with chromosome numbers from 35 to 80 and, in this case, the hybrids could be asymmetric with respect to either parent.

## THE CYTOPLASMIC GENOMES

A further level of complexity relating to the genetic behavior of fusion hybrid is revealed when we consider the fate of cytoplasmic genomes, normally, the mitochondrial and chloroplast genomes, following the fusion. Traits, such as some types of male sterility, disease resistance, and herbicide resistance, are known to be coded onto the genes of these organelles.

Again, competition seems to play a role in elimination of one group of parental chloroplasts or mitochondrial genomes. The reasons for this preferential elimination of genomes are not clearly understood.

Cytoplasmic genomes can be manipulated in a unique fashion by the technique of cybridization. In this variant of protoplast fusion, the nuclear genome of one parental protoplast type is inactivated, e.g., by X-ray irradiation. This may produce a wide range of cytoplasmic variants. In this case, cytoplasmically encoded traits can be manipulated specifically while leaving the majority of the plant's characteristics (encoded in the nucleus) intact. Many researchers see more potential in this technique than in standard protoplast fusion.

## COMMON POTENTIAL OF PROTOPLAST FUSION

This technique has the potential to improve commercially important plant species. However, this does not mean that protoplast fusion will replace the conventional breeding techniques, rather, it will complement breeding programs. The major obstacles are the difficulty in regeneration of many fusion products and the often low fertility of plants that have been recovered.

The crop species that can be regenerated readily from protoplasts are summarized in Table 6.2. Unfortunately, plant regeneration from protoplasts of important crops such as legumes and cereals have not been successful so far.

Protoplast fusion has potential in plant breeding. However, crop improvement is a multimillion dollar commercial enterprise and, in most situations, the technique used is that which produces the end product cheaply, rather than the one which is most scientifically significant or technically clever. Therefore, the possible gains due to fusion of protoplasts must be viewed alongside other ways of achieving the same results. Somatic hybridization through protoplast fusion only makes sense when there is a barrier such as sexual incompatibility. Certain types of incompatibility can be overcome by other means more readily than through protoplast technology. For example, some potentially desirable sexual crosses produce embryos that abort be-

TABLE 6.2. Important Crop Plant Species That Are Amenable to Plant Regeneration from Protoplast

| Alfalfa | P | *Medicago sativa* |
|---------|---|-------------------|
| Asparagus | C | *Asparagus officinalis* |
| Barley | C | *Hordeum vulgare* |
| Carrot | P | *Daucus carota* |
| Clóver | P[1] | *Trifolium* spp. |
| Cucumber | P | *Cucumis sativus* |
| Endive | C | *Cichorium endivum* |
| Lettuce | P | *Lactuca sativa* |
| Maize[2] | C | *Zea mays* |
| Millet | C | *Pennisetum americanum* |
| Orange | P | *Citrus sinensis* |
| Poplar | C | *Populus* spp. |
| Potato | C | *Solanum tuberosum* |
| Rape | P | *Brassica napus* |
| Rice | C | *Oryza sativa* |
| Strawberry | C | *Fragaria chiloensis* |
| Sugar beet | C | *Beta vulgaris* |
| Sunflower | C | *Helianthus annuus* |
| Tobacco | P | *Nicotiana tabacum* |

*Source:* Stafford and Warren, 1991. © John Wiley & Sons Limited. Reproduced with permission.

P = Species will regenerate from whole plant protoplasts
C = Species will only regenerate from suspension culture-derived protoplasts
[1]Ability to regenerate from whole plant protoplast depends on species
[2]Only a few cultivars

fore mature fertile seed can be produced. The technique of embryo rescue (as discussed in Chapter 5) may be used to overcome this problem.

Protoplast isolation and fusion has a place in situations when we are only interested in cell culture and in using the products of these cells for pharmaceutical, medicinal, or food-related products. In this situation, regeneration of fused protoplasts is not going to be a barrier. In fact, success has been achieved with two species. *Euphorbia millii* produces the food color anthocyanin and *Coptis japonica* produces berberine, a pharmaceutical compound. The fusion of the protoplasts of these two plants has resulted in cells which produce both substances. They are grown commercially using bioreactors.

Furthermore, protoplasts can be used in genetic transformation using *Agrobacterium* or by electroporation or microinjection.

Protoplast fusion and gene manipulation are, at present, complementary skills, but in both cases their effective implementation awaits further advances in our understanding of the techniques and the characteristics we wish to manipulate.

Chapter 7

# Cell Culture and Selection
# of Desirable Traits

## SELECTION OF NATURALLY OCCURRING
## VARIANTS IN CULTURE

The original concept of micropropagation was that all the plants originating from a somatic tissue should be genetically identical (i.e., clones) to the donor plant from where the explants were obtained, and no variation was expected. Therefore, when some variants were detected among the regenerents (the "clonal progeny") it was usually dismissed as the "effect of tissue culture" or the "effect of exposure to plant growth regulators." Larkin and Scowcroft (1981) coined the term "somaclonal variation" to describe the observed phenotypic variations in the regenerates in micropropagation experiments.

When it was recognized as a genuine phenomenon, somaclonal variation was considered to be a potential tool to introduce new variants for perennial crops that are asexually propagated (e.g., banana). The case of obtaining thornless blackberry plants as such a variant has been referred to in Chapter 3. Some of these variations occur due to physiological (e.g., prolonged exposure to 2,4-D), biochemical, or even genetic causes. The biochemical and genetic causes may manifest mainly due to minor genetic variations in the cell populations (polysomatic behavior) within the explants used to initiate tissue culture. Thus, if plants are regenerated from these distinct populations of cells, two or more variants among the regenerents would be observed (Jayasankar, 2000). Also, alteration in the methylation status of the cytosine base in the genomic DNA can be brought about by tissue culture, leading to epigenetic variants in the regenerated plants.

Two of the earliest success stories of selection of variants from cell culture are the isolation of tryptophan-accumulating cell lines of car-

85

rot and tobacco (in 1972), and streptomycin resistant (SR1) tobacco plants from callus cultures (in 1973; Dix, 1994; Jayasankar, 2000). The SR1 line of tobacco was perhaps the first case of an in vitro selected line that had a sexually inheritable phenotype for streptomycin resistance (Maliga, Sz. Breznovitis, and Marton, 1973). Subsequently, the resistance was shown to be due to a single base substitution in the 16SrRNA gene in the chloroplast DNA of the variant plants.

## GENERAL SELECTION STRATEGIES

Selection of variants in cell culture can be done using several approaches.

· 1. Positive selection: This is achieved by subjecting a large population of cells in culture (preferably cell suspension or protoplast cultures) to sublethal doses of a selected toxin or antimetabolite. The cells that grow at a normal rate are selected, while the growth of the wild-type cells will be arrested under these conditions.
2. Negative selection: This is done using counter-selective agents to kill the wild-type cells and permit the growth of resistant lines or auxotrophs. Selection of photoautotrophic cells from a heterotrophic callus or cell culture by withholding sucrose from the nutrient medium may also be grouped under this category.
3. Other selection strategies employed include visual (based on differences in pigmentation or type of growth habit—friable versus compact callus), or temperature sensitivity of the regenerates and the elegant strategy for selecting virus resistant tobacco lines (described below).

### Selection for Herbicide Tolerance

Herbicide-tolerant cell lines have been obtained for several species. This includes glyphosate-tolerant cell lines of petunia, carrot, tobacco, and tomato. In all these cases, the tolerant lines had elevated levels of enolpyruvyl shikimate phosphate (EPSP) synthase, which is the target enzyme of the broad-spectrum herbicide, glyphosate. The overexpression of the target genes (for specific enzymes in the case of herbicide tolerance) appears to be the general mechanism by which cells acquire tolerance to herbicides. Thus, alfalfa cell lines that showed tolerance to phosphinothricin herbicide had elevated levels of glutamine synthetase.

Several other examples exist for selection of cell lines tolerant to other herbicides. These include sulfonylurea-tolerant flax and oilseed rape cell lines. Also, maize cell lines tolerant to imazaquin, an imidazolinone herbicide, were selected from embryogenic cell cultures. The whole plants regenerated from such variant lines retained a high degree of tolerance to the herbicide in a sexually heritable fashion.

The more recent developments of genetic engineering provided the option of introducing cloned genes that can confer tolerance to herbicides of choice. This has made the laborious process of searching for variants in vitro rather redundant. Nevertheless, if mutant plants with highly desirable agronomic traits are isolated by cell culture (Dix, 1994), the objections about the use of genetically modified organisms (GMO) as a source of food may be avoided. Therefore, the search for suitable variants in cell culture can still be justified.

## Selection for Disease Resistant Lines

The selection strategy for disease resistance in vitro is by screening for resistance either to the pathogen or to toxins produced by the pathogen. Often, the crude culture filtrates of the pathogen (i.e., the medium in which the pathogen was grown) have been used as the source of toxin for ease of preparation. Thus, Carlson (1973) was able to regenerate tobacco plants resistant to the toxin produced by *Pseudomonas tabaci* (a bacterium that causes the wildfire disease in tobacco) by selecting protoplast-derived cell cultures in the presence of methionine sulphoximine (a toxin produced by the bacterium). Since then, fungal and bacterial toxins have been employed to obtain disease resistant plants of several agronomically important species such as rice, maize, tobacco, and alfalfa (see Table 7.1). A generalized scheme for obtaining disease resistant plants from screening of cell culture is also presented in Figure 7.1.

There are several reports of successful isolation of resistant cell lines without toxin selection. Therefore, in some of the cases, the efficacy of the toxin selection may not be higher than the harnessing of naturally occurring variation in cell culture. Thus, a sorghum line resistant to attack by autumn armyworm has been obtained from callus

TABLE 7.1. A Partial List of Agronomically Important Plant Species for Which Disease Resistant Plant Lines Have Been Obtained by Selection of Variant Cells In Vitro

| Crop | Pathogen | Toxin |
|---|---|---|
| Alfalfa | *Colletotrichum* sp. | Culture filtrate |
| Banana | *Fusarium* sp. | Fusaric acid |
| Coffee | *Colletotrichum* sp. | Partially purified culture filtrate |
| Maize | *Helminthosporium maydis* | T-toxin |
| Oat | *Helminthosporium victoriae* | Victorin |
| Oilseed rape | *Phoma lingam* | Culture filtrate |
| Peach | *Xanthomonas* sp. | Culture filtrate |
| Potato | *Phytophthora infestans* | Culture filtrate |
| Rice | *Xanthomonas oryzae* | Culture filtrate |
| Sugarcane | *Helminthosporium* sp. | Culture filtrate |
| Sugarcane | *Helminthosporium sachari* | Partially purified HS toxin |
| Tobacco | *Pseudomonas tabaci* | Methionine-sulfoximine |
| Tobacco | *Alternaria alternata* | Partially purified toxin |

*Note·* The disease-resistant phenotype of oat, oilseed rape, rice, and tobacco lines listed above were shown to be heritable through sexual propagation. Potato and sugarcane lines were stable through vegetative propagation.

cultures. Such resistance had not been known to occur in sorghum before.

Toyoda et al. (1989) obtained tobacco plants resistant to tobacco mosaic virus (TMV) by an innovative strategy that shows promise to be a method of choice for selecting virus-resistant plants in other plant species too. They cultured TMV-infected cells, and from the callus obtained they isolated cells with both high TMV titre and high growth rates—indicating that these cells were resistant to the TMV. They were able to obtain healthy plants from such selected callus and demonstrated that those plants were resistant to TMV, and that the trait was inherited as a single dominant mutant gene by the sexual progeny.

### *Selection for Amino Acid Accumulators*

Apart from herbicide tolerance and disease resistance, high accumulators of specific amino acids are another phenotype for which se-

Donor plant tissue

▼

Callus cultures

    ▼  Several rounds (3 or more) of selection on
        medium with the toxin

Population of regenerated plants

    ▼  Test for toxin resistance

Putative disease resistant plants

    ▼  Vegetative propagation or selfing

Field planting and testing the progeny

    ▼  Test for disease resistance

Disease resistant plants

    ▼  Vegetative propagation or selfing

Amplified progeny population

    ▼  Further selection of disease resistant plants
        and propagation of stable resistant plants

Use the selected plants for breeding exercises

▼

Improved variety released to the growers

FIGURE 7.1. A generalized scheme for obtaining disease-resistant plants from screening of cell culture.

lection can be performed in vitro. We will focus on the examples from cereals under this category. Lysine is an essential amino acid (i.e., it is one of the amino acids we cannot synthesize on our own—it is obtained exclusively from food intake) that occurs in relatively low abundance in the cereal proteins (Schaeffer and Sharpe, 1990). Thus, a significant proportion of the effort to isolate amino acid-accumulating cell lines focuses on cereals such as rice, wheat, maize, barley, and pearl millet.

In all these cereals, whole plants with higher lysine content in the seed (grain) protein have been isolated. In most cases, the trait was inherited as either a single locus or as two dominant loci (genes), except in rice, where it was shown to be inherited as a recessive trait (Schaeffer and Sharpe, 1990). In rice this characteristic was sometimes inherited along with certain undesirable traits such as infertility and abnormal seed fill. Despite these setbacks, work is underway to utilize these high lysine lines in breeding programs. With further refinements in technology, we may indeed see new cereal crop varieties with high lysine in the grains released to the growers in the not too distant future.

### Selection for Variants for Resistance to Abiotic Stresses, with Special Reference to Salt Tolerance

Abiotic stresses to which plants are subjected include salt and heavy metal ions in the soil and water, oxidative, UV, and temperature (cold and heat) stresses. For plants, any level of these factors that differs significantly (more or less) from that to which they are usually exposed can be considered stressful. The loss of yield due to such stresses can be tremendous. Thus, it would be desirable to obtain crop plants that can tolerate specific stresses, and yet maintain high yield. For example, if a rice variety that can tolerate relatively high levels of salt is developed, most of the coastal land that has high salt content can be brought under rice cultivation. Also, irrigation with brackish water, if not seawater directly, is possible—thus alleviating two problems in one attempt.

Several cases of isolation of cell lines with higher tolerance to salt concentration in the nutrient medium have been reported, e.g., in rice, mustard, and *Nicotiana plumbaginifolia*. The mechanism of tolerance to some of the abiotic stresses may be under the control of some master regulatory genes because some of the resistant cell lines isolated show cross tolerance to multiple stresses. For example, salt-tolerant lines tend to be drought tolerant too. Also, a sugarcane cell line resistant to growth inhibition by the proline analog hydroxyproline showed greater tolerance to polyethylene glycol (PEG) and low temperature stress.

The mutants selected from cell cultures without mutagen treatment (see Chapter 8) should be treated as "variants" until heritability of the trait is established, because some of the variation may be due to physiological or epigenetic alterations. In such cases, the trait is not passed on to the next generation, making it ineffectual for plant breeding. Often the correlation of observed tolerance to stresses in cell cultures might not be manifested in whole plants grown in the field. As previously mentioned, the resistant alleles can be recessive. Further, it should be kept in mind that the plants regenerated from selected cell lines in tissue culture are not directly usable for farm planting. They should undergo several generations of growth and stabilization of the trait in the field, be put into breeding programs, and the trait transferred to desirable genetic backgrounds before the plants become agronomically desirable.

Nevertheless, the possibility that plant breeders can generate crop plants with increased tolerance to multiple stresses (biotic and abiotic), and with improved nutritional quality entirely by selective breeding (with the help of tissue culture and molecular biology) is very high.

# Chapter 8

# In Vitro Mutagenesis

The expression "in vitro mutagenesis" has been used in several fields, e.g., tissue culture and molecular biology. For our discussion, we will use this term to mean "induction of mutation in cell cultures maintained in vitro by the use of chemical or physical mutagens and subsequent establishment of mutant cell lines and/or regeneration of mutant plants." The definition of the term in molecular biology is more focused, where it is meant to describe the introduction of mutations in isolated genes (DNA in vitro—or DNA outside the cell) and reintroduction of the mutated gene into suitable expression systems. Using such strategies, molecular biologists are able to study the functional domains of genes, e.g., using in vitro mutagenesis, scientists have unequivocally proven that ETR1, a putative ethylene receptor protein from *Arabidopsis,* binds the gaseous hormone ethylene (Rodriguez et al., 1999). This interpretation of the term is outside the scope of this chapter.

The naturally occurring mutation rate is quite low—about 1 in 10 million cells in tissue cultures. Therefore, if one intends to isolate desirable mutants, it would be more efficient to induce a higher mutation rate with chemical or physical mutagens. Induced mutations (as opposed to naturally occurring mutations) are of great use for plant breeding, either directly to improve specific traits, or indirectly for cross breeding experiments (Negrutiu, 1990). Cells have evolved an elaborate set of enzymes to counteract the DNA damage. They can repair and maintain DNA integrity, making natural mutations quite rare. The principle of in vitro mutagenesis, therefore, is to devise a scheme by which we can induce DNA lesions in a certain population of cells maintained in vitro and allow these cells to divide rapidly so that the repair mechanism introduces minor errors in the nucleotide sequence of the DNA. As a result, the selected population of cells would have mutations in specific genes, and if whole plants were regenerated from such cells, one would obtain mutant plant lines.

One such approach, *viz,* selection of resistant mutants, has been alluded to in Chapter 7. The selection of resistant mutants is based on the principle of selecting survivors in a large population of cells treated jointly with mutagens and the toxin against which resistance is sought. The mutagen treatment would induce random mutations in the genome, one of which might confer resistance to the toxin that is also present in the selection medium. By culturing the selected cells, one can recover plants resistant to the toxin or the pathogen that produces the toxin.

In vegetatively propagated plants, isolation of "sports" (i.e., naturally occurring mutants) and establishing them as new varieties is a proven practice. In fact, this has been the only way to obtain improved varieties in exclusively vegetatively propagated species such as apple, potato, cultivated rose, etc. According to one estimate, about 35 percent of the 1,440 commercial varieties of rose, 25 percent of the apple varieties, and 45 percent of potato seed acreage in the United States were derived from such sports or spontaneous mutants (Donnini and Sonnino, 1998). The application of in vitro mutagenesis has vast potential for increasing the available genetic variants in the years to come. By the year 2000, over 2,200 mutant varieties of plants (mostly ornamentals) had been released worldwide (FAO/IAEA statistics), including 175 crop plant species with induced mutant varieties (Maluszynski et al., 2000).

The mutagens cause various kinds of DNA damage, such as deletion or duplication of nucleotides, or rearrangements (inversion, translocation) of segments of DNA in the chromosomes. Some of the base pair deletions and substitutions (e.g., exactly three bases of a codon within a gene) may not lead to frameshift mutations and may not result in any change of phenotype.

## TYPES OF MUTAGENS

There are several types of mutagens available for plant breeders. Among these are physical and chemical mutagens and insertional mutagens (including transposons and T(transfer)-DNA of *Agrobacterium* sp.).

## Physical Mutagens

Physical mutagens include ionizing radiation such as X rays, gamma rays, neutrons, and UV radiation.

*X rays* are electromagnetic radiation produced by electrically accelerated electrons in high vacuum. The wavelength of X rays ranges from 0.001 to 10 nm, and as a result, the penetration of such radiation into tissue can be variable—from a few millimeters in the case of long wavelengths, to a few centimeters in the case of short wavelengths.

*Gamma rays* are emitted from radioisotopes during the natural decay of isotopes such as cobalt-60 and cesium-137 (with half lives of 5.3 and 30 years, respectively). They have shorter wavelengths than X rays, and are more penetrating (can enter several centimeters into the tissues). The dosage needs to be determined by experimentation.

*Neutrons* are produced in nuclear reactors or accelerators. The emission of neutrons results in the release of large amounts of energy that can cause mutations. However, the access to a neutron source may be the major limiting factor for its application to in vitro mutagenesis.

*UV radiation* is known to induce both frameshift mutations (deletion or addition of bases *not* in multiples of three) and base pair substitutions. The formation of pyrimidine dimers in response to UV irradiation, followed by the excision repair mechanisms, leads to single-strand gaps in DNA. This is the main mode of induction of mutation by UV radiation.

Mercury arc lamps that emit in the range of 250 to 290 nm wavelength are the source of UV light in laboratories. Among the ionizing radiations listed, UV light is perhaps the cheapest and most readily available source of radiation. However, due to the relatively long wavelength of UV light, it has limited tissue penetration. Hence, it is suitable for inducing mutations in cell or protoplast cultures. Usually protoplasts are placed a few centimeters below the UV lamp in Petri dishes (without the lid) for durations that lead to 10 to 50 percent reduction in colony formation compared to nonmutagenized protoplasts on subsequent culturing.

## Chemical Mutagens

There are several chemicals that act as mutagens (Figure 8.1). One common feature of these chemicals is that they are all highly toxic

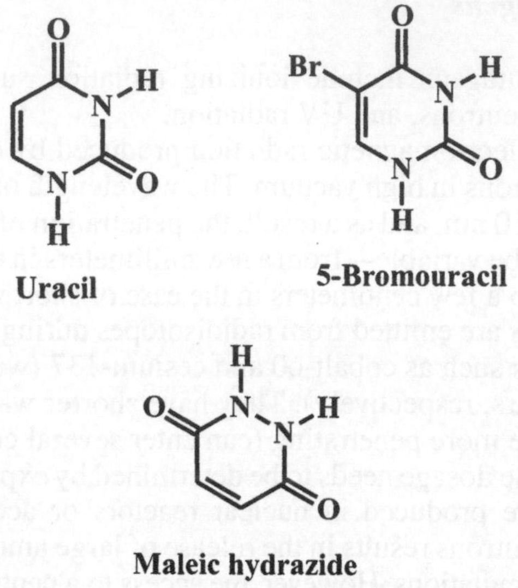

**FIGURE 8.1.** The structure of 5-bromouracil and maleic hydrazide, which are used as mutagens, compared with that of uracil. Due to the close similarity among them, they are referred to as structural analogs.

(they can cause mutations in our body cells that can lead to cancerous growth) and should be handled with extreme caution. They include base analogs, alkylating agents, and other toxic chemicals such as sodium azide.

*Base analogs* are structurally similar to DNA bases, thus, they are easily incorporated into the DNA molecule during its replication at the onset of cell division. The incorporation of promutagenic base analogs into DNA causes aberrant hydrogen bond formation (Negrutiu, 1990). Subsequent base ionization leads to alteration of the bases in the DNA, which is the basis of induction of mutation by the base analogs. The use of such compounds leads to low frequency mutations in plants, hence their use in this field is rather limited.

Some of the compounds in this category include: 5-bromouracil, 5-bromo (or fluoro)-deoxyuridine—both of which are thymine (T) analogs. They can be mistakenly incorporated into the DNA in place of T. In addition, 5-bromouracil can also mispair with guanine (G, instead of with adenine), and this can lead to the substitution of a guanine in place of adenine (A). The end result of such mistaken pairing is the introduction of a G-C pair in place of an A-T pair, leading to a mutation. Also, included under this category is maleic hydrazide (a commercially used selective herbicide that prevents bushy growth in some plants), which is a structural isomer of uracil and can cause sister chromatid exchanges, leading to possible mutations.

*Alkylating agents* are highly reactive compounds that can alkylate phosphate groups as well as purine and pyrimidine bases of DNA. The alkylated bases are subjected to repair which can lead to erroneous repair/substitutions leading to frameshift mutations and strand breaks. There are two categories of compounds under this subclass of mutagens.

1. Alkyl sulfates and sulfonates: examples of compounds under this category include methyl-, ethyl- or propyl-methane sulfonate. Among these, ethyl methane sulfonate (EMS) is commonly used for inducing mutations in plants. EMS is a highly toxic volatile liquid that needs to be pipetted into the cell culture media at a final concentration range of 0.1 to 3 percent v/v.
2. Nitroso compounds include dialkyl nitrosamines, alkyl nitrosoureas, sulphur or nitrogen mustards (e.g., mustard gas), and epoxides. Among these, N-methyl-N-nitrosourea (NMU) and N-ethyl-N-nitrosourea (NEU) are the more commonly used mutagens (at a final concentration range of 0.1 to 10 mM) for plant cell cultures.

### Precautions for Handling the Chemical Mutagens

As mentioned earlier, these chemicals are highly toxic and carcinogenic. They should be handled with extreme caution. The workers should wear appropriate face masks and protective clothing, and if skin contact occurs, the area should be thoroughly washed without delay.

The nitrosoureas are not stable in alkaline solutions, therefore the treatment solution should be the culture medium at pH 5.6 (the usual pH for plant cell culture media). Further, it would be desirable to prepare fresh solutions immediately prior to use.

Proper disposal of the chemicals is also imperative. The nitrosourea-contaminated solutions should be added slowly to excess of 5 percent

NaOH, left in a fumehood prior to flushing down the drain followed by copious amounts of tap water. Contaminated glassware and equipment may be soaked overnight in 5 percent NaOH, and washed again with 5 percent NaOH prior to thorough rinsing (Dix, 1999). EMS-contaminated solutions may be added to a large excess of 3M KOH in 95 percent ethanol, heated under reflux, and repeatedly stirred for two hours prior to discarding.

## DETERMINING THE TYPE AND SUITABLE CONCENTRATION OF MUTAGENS

Determining the type and suitable concentration of mutagens should be done based on data from preliminary experiments. The selected compound and the concentration used should result in a 10 to 50 percent reduction of colony formation from treated cell or protoplast culture on subsequent plating. The treatment is usually given for 60 minutes, followed by several rinses with aseptic culture medium. The treated cells or protoplasts can then be plated on appropriate selection medium for isolation of the mutant cell lines.

## THE CHOICE OF PLANT TISSUES FOR IN VITRO MUTAGENESIS

In general, seeds are the most common plant materials used for mutation breeding, perhaps due to their ease of handling and recovering plants subsequent to the treatment. However, cell cultures or protoplast cultures are employed for in vitro mutagenesis. It would be desirable to use highly regenerative cell lines if plants are to be regenerated after mutagens treatment. If protoplasts are employed for the experiment, it would be easier to use them a couple of days after isolation, so that they would have regenerated a portion of their cell wall and will not burst due to excessive handling. Shoot apical meristems may also be used for such experiments, but the resultant plants tend to be chimeric in nature and several generations of progeny will have to be screened before a stable mutant line is obtained. Table 8.1 gives some examples of mutant plants obtained by in vitro mutagenesis and the tissue used.

TABLE 8.1. Some Examples of Mutant Plants Obtained by In Vitro Mutagenesis

| Plant | Type of Mutagen | Tissue Used | Phenotype of Mutant Lines |
|-------|-----------------|-------------|---------------------------|
| Potato | Gamma ray | Shoot cultures | Shallow eye, altered tuber shape, size, and skin color |
| Chrysanthemum | Gamma ray | Shoot cultures | Altered flower shape, color, floret size |
| Carnation | Gamma ray | Shoot cultures | Altered flower color, shape and leaf size |
| Carnation | X rays | Nodal cuttings in vitro | Altered flower color |

## Insertional Mutagenesis

Transposon and T-DNA insertion into the genomes of plants causes "insertional mutations." The identities of the affected genes by chemical and physical mutagens are not readily obtained compared to the ease with which such information can be obtained for insertional mutants. The sequences of the transposons (jumping genes) or the T-DNA segments inserted into the host plant genome are known. Therefore, these insertions serve as markers for the chromosomal locations of the mutated genes, and also facilitate the cloning of the wild-type gene. Visual selection of the mutant plants is feasible by the altered phenotype from the population of transgenic plants.

This category of mutagenesis was not fully exploited until the mid-1980s, when methods for obtaining transgenic plants became firmly established. However, the existence of transposons and their ability to cause mutations in certain plants (e.g., maize and the snapdragon) was known for much longer (first reported by Barbara McClintock (1951) in the late 1940s).

The molecular genetic analysis of flower development using *Arabidopsis thaliana* and *Antirrhinum majus* was aided primarily by the development of mutants obtained by transposon or T-DNA insertional mutagenesis. Thus, transposon mutagenesis helped in identifying and cloning the *Floricaula* gene of *Antirrhinum* (Coen et al., 1990). Similarly, the identification and functional characterization of the *Agamous* gene of *Arabidopsis* was carried out with the help of T-DNA insertional mutagenesis (Yanofsky et al., 1990). There have been several mutants isolated with variations of insertional mutagenesis (e.g., enhancer trap, and gene trap methods) during the past few years (Sundaresan et al., 1995; Liljegren et al., 2000). Some of these mutants are used for basic research and functional analysis of selected genes, while others are usable as new varieties of ornamental plants. With technological advances, an increasing number of useful mutants will be released in the near future.

Figure 8.2 summarizes a scheme for obtaining induced mutants by in vitro mutagenesis as it stands at present time.

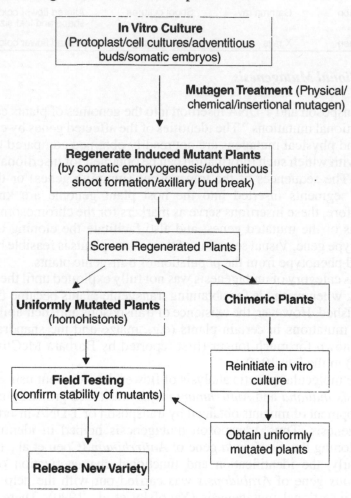

FIGURE 8.2. A scheme for obtaining induced mutants by in vitro mutagenesis.

Chapter 9

# The Origin, Nature, and Significance of Variation in Tissue Culture

## INTRODUCTION

The physical appearance of a plant is determined by its genetic makeup and the environment in which it is growing. Every plant species possesses a unique set of genetic information encoded in the genetic material deoxyribonucleic acid (DNA) as discrete hereditary units called genes. Most of these genes are assembled into nuclear organelles, the chromosomes, and the rest are contained within mitochondria and chloroplasts. The nuclear DNA content varies within species, and different species are distinguished by having a characteristic basic chromosome number. In nature, vegetative cells normally possess twice the basic number of chromosomes, referred to as diploid (2n), but the existence of plants with other ploidy levels such as haploid (n), triploid (3n), and tetraploid (4n) causes considerable genetic variation in many natural populations. Genetic variations due to structural changes in chromosomes have also been noted.

It is now well documented that a number of external factors, including the process of tissue culture, can induce heritable genetic changes in plants. After considerable research and reflection, Larkin and Scowcroft (1981) adopted the term "somaclonal variation" to describe the genetic variation occurring in in vitro cultured cells, tissues and plants. They have suggested the potential of tissue culture for the induction of useful and stable variations that could be exploited for crop improvement. In this chapter the basis and the significance of somaclonal variation will be discussed.

## THE BASIS OF SOMACLONAL VARIATIONS

There are several theories to account for the origin of somaclonal variation in plant tissue culture. They can be grouped into two categories.

1. Variations originating from the genetic heterogeneity of somatic cells of source plant
2. Variations due to structural alterations of DNA and chromosomes caused by tissue culture

### Genetic Variation Arising from Source Plant

The component cells of the primary explants used to initiate cultures are likely to be heterogeneous with respect to the state of differentiation, ploidy level, and age. These explant-related factors are shown to affect the genetic makeup of the cells produced in the culture. For instance, all the cell types in an explant can be induced to divide in culture, and it is highly likely that the callus arising from such a group of cells with diverse genetic makeup will inevitably lead to a mixed population of cells. Plants regenerated from such genetically mosaic callus will undoubtedly be of different genetic makeup, depending on the cell types from which the plants are originated. Genetic mosaicness seems to occur commonly in polyploid plants rather than in diploids or haploids.

### Genetic Variation Arising During Culture

Although a significant degree of genetic variability can be traced to, at least in polyploid species, the genetically heterogeneous cell types of explant, there is substantial evidence to indicate that much of the variability observed in regenerated plants stems from the culture process itself. Aneuploids, polyploids, or cells with structurally altered chromosomes may arise in culture. Many differentiated cells, when induced to divide in culture undergo endoduplication of chromosomes, resulting in the production of tetraploid or octaploid cells with distinct phenotypes.

Occurrence of multipolar spindles due to failure of spindle formation during cell division has been attributed to the production of cells with unusual ploidy levels. Absence of spindle formation during mitosis results in the appearance of cells with doubled chromosome number while the formation of multipolar spindles on chromosomes lagging at anaphase causes the development of cell lines with haploid, triploid or other uneven ploidy status. All these phenomena are observed in tissue culture of various plant species (Bhojwani, 1990).

Many recent studies indicate that cryptic structural modification of individual chromosomes is more likely to cause somaclonal variation than modification induced by ploidy changes in many tissue-cultured plants. Chromosomal changes occurring during tissue culture include transposition of mobile genetic elements called transposons, chromosome breakage, and repositioning of chromosome segments.

## CAUSES OF SOMACLONAL VARIATIONS

Several mechanisms have been advanced to explain the genetic variability occurring in tissue culture. The most possible causes are:

1. *Reduced regulatory control of mitotic events in culture.* The ploidy status of plants regenerated from callus, cell suspension or protoplast cultures of certain species differ significantly despite the fact that the cultures are originating from a highly homogenous genetic background. This indicates the lack of tight regulation of cell cycle-related controls during cell proliferation in culture.

2. *Use of growth regulators.* Plant growth regulators, particularly synthetic auxins such as 2,4-dichlorophenoxyacetic acid (2,4–D), are considered the major cause of genetic variability in culture. Interestingly, cytokinins at low concentrations have been shown to reduce the range of ploidy in culture. Also, low levels of both auxins and cytokinins appear to preferentially activate the division of cytologically stable meristematic cells, enabling the regeneration of genetically uniform plantlets.

3. *Other medium components.* Some of the mineral nutrients influence the establishment of genetic variability in culture. For instance, by altering the levels of phosphate and nitrogen as well as the form of nitrogen in the medium, the genetic composition (ploidy level) of the cultured cells can be controlled to a considerable extent. A marked increase in chromosome breakage has

been observed in plant cell cultures grown with different levels of magnesium or manganese.

4. *Culture conditions.* Some culture conditions, like incubation temperature above 35° C and long duration of culture, have been implicated in inducing genetic variability in regenerated plants.

5. *Inherent genomic instability.* Some of the recent molecular studies indicate the existence of certain regions of genome that are more susceptible to tissue culture-induced structural alterations. The reason for the increased susceptibility of these genomic loci, called "hot spots," is not known.

### Cytoplasmic Genome and Somaclonal Variation

There are only a few examples to suggest that the tissue culture process itself causes heritable genetic mutation in chloroplast or mitrochondrial genomes. Mutations affecting male sterility have been observed in maize and tomato cultures and it is highly likely that mitochondrial DNA has been affected in both species. In tobacco, some distinct changes in mitochondrial DNA sequence occurred in suspension cultured cells and these DNA sequence alterations have been implicated in the phenotype variability observed in regenerated plants.

### Causes of Epigenetic Variation in Tissue Culture

An important aspect to be highlighted at this point is the occurrence of epigenetic variation in tissue culture. Frequently, any culture-induced changes, which can be stable but not heritable, have been considered as epigenetic variations. However, a greater understanding of genetic and epigenetic alterations in culture in the recent past led to a clear distinction between these two types of variation. For instance, genetic mutations occur randomly and at a much lower rate than epigenetic variations. Genetic changes are usually stable and heritable. Epigenetic variations may also lead to stable traits; however, reversal can occur at high rates under non-selective conditions. Epigenetic traits are often transmitted stably through mitosis but rarely through meiosis, and the level of induction of epigenetic traits is directly related to the selection pressure experienced by the cells. Epigenetic changes are generally assumed to reflect alteration in expression rather than the information content of genes. The epigenetic variations observed in cultured cells or regenerated plants are mainly due to three cellular events:

1. Gene amplification
2. DNA methylation
3. Increased activity of transposable elements

As with many animal systems, some plant cells in culture possess the ability to grow and multiply when exposed to certain toxins, drugs, or other growth limiting conditions. It is not clear how these cells acquire resistance to these conditions, but it is suggested that gene amplification could be a possible mechanism conferring resistance. Plant cells resistant to antimetabolites often exhibit an increased amount or activity of target enzymes as exemplified by the profile of glutamine synthetase in phosphinothricin-resistant alfalfa cell lines. Epigenetic variant cells need to be subjected to constant selection pressure to maintain resistance traits.

In plants, nearly 25 percent of the genome can be methylated at cytosine residues but the significance of this cytosize methylation is not apparent. It has been suggested that methylation (and demethylation) of DNA is one of the ways of controlling transcriptional activity and this process has been affected by tissue culture process. The nonheritable genetic variability observed in many tissue culture systems could thus be attributed to tissue culture–induced methylation or demethylation of DNA. Some recent studies clearly demonstrated a role for DNA methylation in the regulation of organogenesis in *Petunia* and *Paulownia fortunei* leaf culture (Prakash and Kumar, 1997). Table 9.1

TABLE 9.1. Examples of Modification in DNA Methylation Pattern Observed in Tissue Culture

| Plant species | Type of culture | Regenerates obtained | Observed changes in methylation pattern |
|---|---|---|---|
| Petunia | Leaf culture | Plantlets | Inhibition of shoot induction involves hypomethylation of genomic DNA |
| Soybean | Cell suspension | Plantlets | Reduced methylation in ribosomal DNA |
| Carrot | Cell suspension | Somatic embryos | Changes in genomic DNA methylation pattern during embryogenesis |

presents examples of modification in DNA methylation pattern observed in tissue culture.

The observations related to the existence of genetic elements (transposons) capable of moving around the genome by Barbara McClintock in the 1940s provided a major breakthrough in our understanding of gene regulation (McClintock, 1984). A remarkable increase in the mobility of these transposons as well as the much later discovered retrotransopons, another group of mobile genetic elements, was observed in different species in tissue culture (Kumar and Bennetzen, 1999). For instance, in a study reported by Groose and Bingham (1986), nearly 25 percent of the *Medicago sativa* regenerates showed flower color variation, which is thought to be caused by transposon activity triggered by tissue culture. Many later studies concluded that the activity of transposons and retrotransposons induced by tissue culture could be responsible for some of the genetic and epigenetic variability observed in culture.

## USE OF SOMACLONAL VARIATION IN BREEDING

It is apparent that plant tissue culture, especially cell and callus culture, produces considerable genetic variability which can be exploited for many useful applications including plant breeding. Although the variability in cultures was identified as early as the 1950s by White, Gautheret and Street, the pioneers of tissue culture, the practical utility of somaclonal variation in crop improvement was not realized for a long time (Gautheret, 1985). To date, a considerable number of useful somaclonal variants have been generated from different plant species and some of them are listed in Table 9.2.

TABLE 9.2. Some Examples of Somaclonal Variation Observed in In Vitro Regenerated Plants

| Plant species | Type of variation |
| --- | --- |
| Cavendish banana | *Fusarium* wilt resistance |
| Wheat | High yield |
| *Lathyrus sativus* | Reduced neurotoxin ODAP |
| Potato | Reduced browning after peeling |
| Potato | Mildew resistance |
| Blackberry | Thornless cultivar |

| Plant species | Type of variation |
| --- | --- |
| Flax | Salt and heat tolerance |
| Celery | *Fusarium* wilt resistance |
| Celery | Insect resistance *(Spodoptera exigua)* |
| Sugarcane | Resistance to *Helminthosporium sacchari* |
| Tobacco | Mosaic virus resistance |
| Tomato | *Fusarium* wilt resistance |
| Tomato | High solid content |
| Indian mustard | High yield and shattering resistance |
| *Capsicum* | Yellow fruit |
| *Capsicum* | Early maturation |
| Rice | Submergence tolerance |
| Rice | Sheath blight resistance |

The following is a brief outline of a general scheme for the creation of somaclonal.

1. Callus-based method

- Identify a suitable explant which should be profusely callusing.
- Test a number of genotypes for callus production.
- Establish an actively growing callus culture and develop a plant regeneration protocol.
- Subject callus to mutagen (chemical or physical) treatment.
- Expose callus to the desired selective agent. The selective agent could be toxins (culture filtrate of pathogenic fungi, bacteria, or a chemical analog of a natural compound), herbicides, or environmental stresses such as high temperature, cold, high osmoticum, salinity, etc.
- Subculture the callus on a medium containing increasing levels of selection agent.
- Identify a concentration/level of selection agent that kills most of the callus cells but allows some to grow continuously.
- Continue the selection pressure and regenerate whole plant.
- Test for whole plant resistance/tolerance at field level.

2. Cell or protoplast culture-based method

- Use callus for the preparation of cell suspension culture.
- Prepare protoplasts, if desired, from cell suspension or any other desierable plant tissue.

- Standardized cell or protoplast culture method.
- Establish a plant regeneration method from cells or protoplasts.
- Subject cells or protoplasts to mutagen treatment.
- Place cells or protoplasts on a suitable growth medium and follow the steps used for callus-based method for selection and recovery of plants with desirable traits.

## PREVENTION OF SOMACLONAL VARIATION

Although somaclonal variation provides a novel source of variability for crop improvement, in most applications of plant tissue culture, maintenance of clonal stability is a major consideration. From the preceding discussion, it is obvious that the use of preexisting meristem as explant, avoiding the use of strong plant growth regulators such as 2,4–D, a short duration of culture, as well as elimination of other stress conditions likely to trigger transposon and retrotransposon activity may help maintain the clonal fidelity of the regenerated plants. Production of genetically uniform planting material is an essential requirement for the micropropagation industry involved in clonal propagation.

### Techniques for Identifying Somaclonal Variation

The power and precision of molecular methods provide valuable tools for the reliable identification of somaclonal variants in a population. With the recent advances in polymerase chain reaction (PCR) and other DNA-based methodologies, it is possible to identify individuals that differ genetically, even by only a single base pair.

For the identification of somaclonal variants, two different DNA-based molecular methods seem to be appropriate. These methods are based on the techniques of DNA hybridization and PCR.

Hybridization methods rely on the use of specific DNA probes to identify the genetic variability of the plant material. Among the different hybridization methods, Restriction Fragment Length Polymorphism (RFLP) has been used extensively for the identification of genetic variants. This technique involves fragmentation of genomic DNA with various restriction enzymes, followed by the identification of RFLP profile using highly specific DNA probes to establish the genetic uniformity of the plant material.

PCR methods, on the other hand, possess several advantages over hybridization-based methods. These include sensitivity, speed, and specificity. Random Amplified Polymorphic DNA (RAPD) and Amplified Fragmented Length Polymorphism (AFLP) are the two most suitable methodologies for the identification of somaclonal variation in plants (Henry, 1997). These techniques are further described in Chapter 14.

Chapter 10

# Cryopreservation and Plant Breeding

## *INTRODUCTION*

Since the beginning of the last century, various scientific explorations of natural vegetation have led to the identification and detailed description of major geographic centers of cultivated plants and their wild relatives. These vast natural gene pools provided valuable resources to develop our modern high yielding cultivars. For instance, many agronomically useful traits, such as disease resistance and environmental stress tolerance of the present day cultivated varieties, were derived from the wild relatives which acquired these attributes through long exposure to various environmental stresses and pathogens and by introgressing with other wild species. Unfortunately, these natural repositories of germplasm are fast dwindling due to neglect and deliberate destruction for urban development. To stem this decline, and to preserve these reservoirs of genetic diversity, many conservation programs are now being undertaken. These programs rely on various strategies for long-term storage of seeds and other propagules under certain conditions that ensure minimum risk of genetic damage. In the past, a great deal of interest in the application of in vitro methods for the conservation of genetic resources has been expressed. In particular, in vitro storage of plant materials under low temperature, commonly called cyropreservation, proved to be a very valuable tool for the conservation of plant germplasm (Kartha, 1985; Bajaj, 1995).

## *THEORY AND TECHNOLOGY*

By definition, cyropreservation is the nonlethal storage of biological tissues at low or ultra-low temperatures without losing the viability and genetic integrity of the material preserved. The retention of viability and genetic integrity is crucial, as plants need to be regenerated from the preserved material which is used for breeding programs.

The technical complexity of cyropreservation of plant material results from the differential sensitivity of various plant parts and propagules to low temperature and water deficit associated with the cryogenic protocols. Seeds and pollen grains of many crop species tolerate desiccation and ultra-low temperatures and can therefore be preserved for extended periods without any negative effect on cell viability. However, such extremes of environmental conditions are lethal for many other propagules. Thus, amelioration of injuries caused by freezing and desiccation remains the major challenge associated with the low temperature storage of delicate tissues. This chapter summarizes recent technological advancements in the area of cryopreservation and presents various practical considerations for successful implementation of cyropreservation of germplasm, particularly for plant breeding.

## Choice of Plant Material

Plant tissues that faithfully regenerate the clone upon culture would be the ideal material for cryopreservation. Now that procedures for regenerating clonal plants from meristem and shoot tip explants have been established for many plant species, it is only logical to use shoot tips and buds for germplasm conservation.

Attempts to cryopreserve seeds have had limited success. However, there are numerous reports of successful recovery of clonal plants from cryopreserved shoot tips and meristems from various plant species, including those producing recalcitrant seeds (Bajaj, 1995). It appears that cryopreservation of callus and suspension-cultured cells is undesirable due to the high incidence of somaclonal variation, arising from de novo meristem formation, and poor viability of cells under freezing conditions. Interestingly, cryopreserved embryogenic lines retained their embryogenic potential and the plants regenerated from those embryogenic lines showed parental attributes without any significant variation (Kartha et al., 1988; Dodds, 1991).

Certain prefreeze conditions seem to have a remarkable effect in improving the recovery of viability of cryopreserved tissues. Prefreeze conditioning included cold-hardening treatments, and application of medium additives such as trehalose, abscisic acid, proline, glycinebetaine, etc., which are known to improve stress tolerance in many plant species. Various aspects of these prefreeze conditioning methods are discussed as follows.

## CRYOPRESERVATION PROTOCOLS FOR COLD-HARDY AND NON-COLD-HARDY SPECIES

There are distinct differences in cryopreservation procedures employed for cold-hardy and non-cold-hardy species. This is largely due to the difference in the extent of natural cold acclimation between different species. The procedures employed for the cryopreservation of cold-hardy plants, mostly the temperate species that can tolerate –25°C or less, exploit the natural ability of these plants to tolerate freezing stress. Significant increase in freezing stress tolerance and the viability of cryopreserved tissues of cold-hardy plants can be achieved by cooling them slowly at a rate of 2°C per hour to temperatures between –30°C and –50°C before exposing them to liquid nitrogen.

Practically all the non-cold-hardy plants require some form of cryoprotective treatment for their survival at ultra low temperatures. As with cold-hardy species, both slow cooling and rapid freezing by direct exposure to liquid nitrogen can also be used for non-cold-hardy plants.

### Isolation of Shoot Tips and Axillary Buds

Shoot tips and axillary buds, measuring about 0.3 to 1 mm in length, isolated from field-grown or in vitro plants can be used for cryopreservation. Usually these tissues will be soaked in 0.5 to 5 percent dimethyl sulphoxide (DMSO) to enhance survival. For most plants, terminal shoot apex and axillary buds from the top position of the shoot showed a significantly higher survival rate than those collected from the lower part of the shoot system. However, such positional effect on meristem survival was not evident in several plants from the Solanaceae family (e.g., potato).

### Prefreeze Cryoprotective Treatment

### Stock Plant Conditioning

Little is known about stock plant conditioning that enhances the viability of cryopreserved tissues. In general, the survival of cryopreserved tissue is determined, to a considerable extent, by its cold-hardiness. Also, evidence suggests that natural cold-acclimatization prior to freezing minimizes injury of cryopreserved tissues.

### Chemical Cryoprotection of Shoot Tips

Culturing of shoot tips and meristems of both cold-hardy and non-cold-hardy species in the presence of cryoprotectants prior to freezing was found to increase their viability. Two types of cryoprotectants are commonly used for plant material. The permeating cryoprotectants such as DMSO, ethylene glycol, and propylene glycol are the most effective ones and they exert their protective effects through colligative properties. The other nonpermeating or slowly permeating compounds like mannitol, glycerol, polyelthylene glycol, sucrose, and proline confer protection largely by osmotic dehydration. Nonpermeating cryoprotectants were proved to be more effective if combined with permeating cryoprotectants.

The type of suitable cryoprotectant and the required concentration and the approximate duration of exposure varies with species and should be determined by empirical testing. In most cases, non-permeating or slowly permeating cryoprotectants were used at concentrations of 1-2 M, and were incorporated in the medium in a stepwise fashion or as single step additions. Generally, tissues will be incubated for one to three hours in cryoprotectants, often at low temperatures to minimize toxicity, prior to freezing.

### Vitrification

Survival of cryopreserved tissues can be significantly improved by avoiding the ice formation in tissues during freezing. This can be achieved through vitrification, a condition in which water solidifies into an amorphous glassy state instead of forming ice. In cryopreservation, tissues are vitrified by exposing them to highly concentrated osmotically active cryoprotectants and appropriate cooling conditions. Although there are clear advantages in using vitrification to improve tissue viability, this process often causes damaging osmotic effects during the tissue recovery stage.

## Encapsulation and Tissue Desiccation

During encapsulation and tissue desiccation tissues are embedded in calcium alginate beads containing high sucrose. These tissue-embedded beads will be desiccated under sterile conditions, which enable the tissue to withstand freezing conditions without any other cryoprotectants.

## Freezing

The method by which tissues are exposed to ultra-low temperatures can be an important factor in determining the survival of cryopreserved tissues. Samples may be rapidly frozen by direct immersion in liquid nitrogen or cooled gradually at a controlled rate to about –35 to –40°C and held at that temperature for a set time before transfer to liquid nitrogen. Cooling rates of about 0.25 to 1°C per min give optimal survival for plant tissues. Cooling at this rate maintains adequate cell water potential equilibrium and thereby reduces cell injury. Too slow cooling, however, adversely affects cell viability due to many undesirable macromolecular interactions and changes in electrolyte concentrations.

In rapid cooling of tissues, intracellular ice formation, which is often lethal, is inevitable. Intracellular ice formation, therefore, must be minimized to a noninjurious level to retain the viability of the cryopreserved tissues. A successful cryoproservation procedure, whether involving a rapid freezing or a gradual cooling and subsequent freezing step, relies on appropriate prefreeze preparation of tissues, such as vitrification or other suitable cryoprotectant treatments to preserve postfreeze viability of the cells. For instance, with a well standardized procedure involving a direct rapid freezing step, Kumu, Harada, and Yakuwa (1983) reported a 100 percent survival rate with *Asparagus officinalis* shoot tips. Table 10.1 summarizes the examples of successful cryopreservation of meristem tissues of some economically important plants.

TABLE 10.1. Examples of Successful Cryopreservation of Meristematic Tissues of Some Economically Important Plants

| Plant species | Specimen | Specimen preparation and freezing method | Viability, growth and morphogenic response |
|---|---|---|---|
| Asparagus | Shoot tips | 4% DMSO and 3% glucose for 3 days preculture, slow freezing at -40°, then to LN | 100% survival complete plant regeneration |
| Apple | Shoot tips | Hardened for 20 days at -3°C | 100% survival, about 75% plant regeneration |
| Banana | Embryonic cells | Preculture on 6% mannitol for 2-7 days, slow freezing to -40°C, then rapid freezing in LN | 50% survival and plant regeneration |
| Citrus | Somatic embryos | Rapid freezing in LN | 90% survival and plant regeneration |
| | | Slow freezing to -40°C, then rapid freezing in LN | 30% survival and plant regeneration |
| Cassava | Shoot tips | 10% glycerol and 5% sucrose, rapid freezing in LN | 13% survival and plant regeneration |
| Chick pea | Shoot tips | 24 hour preculture in 4% DMSO, slow freezing to -40°C, then rapid freezing in LN | 40% survivial and plant regeneration |
| Pea | Shoot tips | 10% glycerol and 10% sucrose for 15 minutes, then rapid freezing in LN | 100% survival, 60% shoot regeneration |
| Potato | Tuber sprouts | 2-7% DMSO for 2 days, then rapid freezing in LN | 10-20% survival and plant regeneration |
| Rice | Pollen embryos | 7% DMSO, 5% glycerol and 5% sucrose, then rapid freezing in LN | 21% survival and plant regeneration |
| Sugarcane | Callus | 10% DMSO and 0 5 M sorbitol, slow freezing to -40°C, then rapid freezing in LN | 97% survival and plant regeneration |

*Note:* Samples were preserved in liquid nitrogen (LN).

## STORAGE AND THAWING

Frozen tissues are usually stored at −196°C in liquid nitrogen. At this temperature, due to lack of liquid water, chemical reactions hardly occur inside the cells. There is no clear indication to suggest that temperatures lower than −196°C provide any additional benefits for preservation. Indeed, the viability of *Cornus sericea* cells was lost when they were cooled from −196°C to −269°C (Guy et al., 1986).

Little is known about the influence of temperature and the rate at which tissues are thawed on cell viability of cryopreserved tissues. Thawing is critical in vitrified tissues because ice crystallization can occur during this process. In general, rapid warming, which avoids ice recrystallization, retains viability of shoot tips and plant cells preserved with gradual stepwise and rapid-freezing procedures. Warming is achieved by immersing the tissue directly into a sterile water bath set at 35 to 40°C for a few seconds. Excessively dehydrated tissues of some species require manipulation of medium osmoticum level to reduce cell injury during thawing and early recovery.

## Recovery of Cryopreserved Tissues

Survival of cryopreserved tissues is aided considerably by suitable post-thaw treatments to remove the cryoprotectants and by the provision of adequate conditions for rapid recovery of cell growth and multiplication. Cryoprotectants can be removed rather quickly by incubating the tissue in an appropriate liquid medium at low temperature (0°C) but this approach is not suited for all species. Slow release of cryoprotectants from tissues can be achieved by culturing them on semisolid medium with frequent reculturing to fresh medium. For some species, such as sugarcane, rapid removal of cryoprotectants by incubating at higher temperatures (about 20 to 22°C) is more beneficial than 0°C.

To spur cell growth and enhance survival, presence of growth regulators in the post-thaw culture medium is often required. However, care must be exercised to minimize callus formation in order to reduce somaclonal variation, particularly if the material is intended for breeding programs.

## Viability Assays

An integral part of cryopreservation is the use of viability assays to test the effectiveness of cryogenic protocols employed. Various parameters can be used to assess the viability of cryopreserved tissues. Generally, membrane integrity and metabolic activity are regarded as useful indicators of viability, and the assays to test these parameters can be performed relatively easily and rapidly. However, cell/tissue growth is the most reliable criterion for viability, and for this reason, all assays should be compared with growth analysis.

Membrane integrity can be assessed by examining the exclusion of certain dyes from the cells, reduction of tetrazolium salts, accumulation of fluorescein in the cells, and the electrical conductivity of cell leachate. Similarly, assaying the activity of various enzymes, particularly those involved in respiration and biosynthesis of macromolecules and the level of different metabolites is useful for estimating cell viability. However, at this juncture, it must be noted that all the viability assays may not be readily applicable to all the different types of plant materials. For example, assessing membrane integrity could be a suitable approach to establish the viability of protoplasts, while resuming growth under appropriate growth conditions may be more suited for cryopreserved shoot cuttings.

Nondestructive viability assays utilizing nuclear magnetic resonance (NMR) and infrared spectroscopy may be a preferable alternative to other methods as repeated assessment of the same sample can be performed over a period of time without destroying the tissue. Nonetheless, application of these techniques is not popular due to expensive instrumentation.

## *EQUIPMENT FOR CRYOPRESERVATION*

Almost all the biological materials can be cryopreserved using a simple freezing chamber that can be cooled by a solvent system such as liquid nitrogen. However, as we now know, the recovery of a viable tissue is largely determined by our ability to control various stages of freezing and thawing to minimize cell injury. A programmable freezer with a liquid nitrogen-cooled sample chamber would therefore be an essential requirement to ensure success. Provision of liquid nitrogen, low temperature storage space, safety equipment, cryovials, low temperature storage boxes, pump, and a water bath set at 40 to 50°C for thawing samples are other requirements for performing cryopreservation procedures.

## *PRACTICAL ISSUES AND STRATEGIES FOR IMPROVED CRYOPROTECTION*

Although the concept of cryopreservation is inherently simple, the challenge posed by the diversity of plant tissues that can be preserved and our limited knowledge about cryogenics make this technique highly complex (Dodds, 1991; Bajaj 1995). Considerable progress has been made in the recent past to identify the optimal conditions for cryopreservation of different plant tissues. However, we are unsure of the genetic integrity of samples preserved for extended periods. Several bacterial studies suggest that there will be fewer DNA lesions in cells kept under low temperature than with other storage conditions. Thus, cryogenic procedures may not be mutagenic per se. However, realistically, the effect of physical stress experienced by the genetic material during freezing and thawing could possibly result in at least some minor DNA damage. This freezing injury may be more evident in the tropical species than in the cold-acclimated temperate plants.

It appears that some other issues pertinent to cryopreservation still remain unanswered. These issues are:

1. How do cryoprotectants reduce cell damage?
2. What cellular mechanisms confer cold tolerance to temperate species?
3. How does a cell repair the damage that occurs during freezing and thawing?

We do not know exactly how cryprotectants operate at the sub-cellular level. It is known that freezing causes damage at multiple sites in the cells, and the available evidence suggests that the application of different types of cryoprotectants (Table 10.2) incrementally during cooling minimizes damage.

TABLE 10.2. Cryoprotective Ability of Certain Compounds Commonly Used in Cryopreservation

| Low protection | Medium protection | High protection |
|---|---|---|
| Acetyl glycine | Acetyl choline | Betaine |
| Dimethyl acetamide | Diamethyl urea | Diamethyl sulfoxide |
| Glucosamine | Glutamic acid | Ethylene glycol |
| Mannitol | Hydroxyproline | Glucose |
| | Methyl acetamide | Glyceraldohyde |
| | | Glycerol |
| | | Sorbitol |
| | | Sucrose |

Fortunately, intensive research in the area of plant cold-hardiness led to the identification of several mechanisms of freezing tolerance in temperate plants. Significant findings in this area include rapid accumulation of osmotically active compounds such as simple sugars, proline, and glycinebetaine, and compounds that protect nucleic acids and proteins upon exposure to freezing conditions such as spermidine, and spermine. These compounds, particularly certain sugars such as trehalose, mannitol, and amino acids, were successfully used as cryoprotectants in many biological systems.

We have little knowledge about the mechanism by which plants recover from the damage caused by freezing and rapid thawing. There are very efficient mechanisms for repairing nucleic acid damages, and their involvement in retaining the viability of cryopreserved cells cannot be discounted. However, the presence of such sophisticated repair mechanisms for other macromolecules is largely unknown.

Clearly, more effort is needed to develop cryogenic protocols for specific genotypes. Similarly, more innovation is required in the areas of storage, packing, and shipping of cryopreserved materials, as an error at any stage of cryopreservation can lead to lethal injury to the material. A better understanding of membrane biology and cold-hardiness, as well as the rapid advancement in genetic engineering, would undoubtedly help refine the existing procedures and develop novel methods for cryopreservation of plant germplasm.

# Chapter 11

# In Vitro Micrografting

Productivity of plants, in particular tree crops, is affected by infection from viruses and related pathogens. Declines of vigor, yield, and quality are attributed to these disease agents. Severe infections have resulted in the exclusion of some cultivars from commercial usage. Fortunately, there are some methods available to recover pathogen-free plants. For example, thermotherapy is used to provide bud wood that is free of certain viruses, shoot meristem culture is used successfully with many herbaceous plants but its application to tree crops is not universally successful. However, shoot-tip grafting in vitro can be used as a means of elimination of pathogens in trees. Moreover, the plants obtained by this method bypass the juvenile state of a tree (Murashige et al., 1972).

## *DEFINITION OF MICROGRAFTING*

Micrografting involves grafting an apex taken from a mother plant onto (a) a young greenhouse or nursery-grown plant in accordance with accepted grafting techniques (in vivo micrografting), or (b) a decapitated young plant grown from a seedling under aseptic conditions, or a microcutting obtained from in vitro vegetative multiplication (in vitro micrografting). In this chapter the technique of in vitro micrografting is discussed.

### *Initial Technique*

During micrografting, the top of a bud from a tree is dissected under aseptic conditions, and only the apical dome is taken. Generally, it is the meristem associated with the leaf primordia which is isolated; the isolated fragment can be called the apex. Sometimes the bud itself, once rid of the few outlying scales, is used. The organ isolated from the adult tree is grafted onto a rootstock, which is a young plant or herbaceous cutting, using the in vitro micrografting techniques. The stock is usually a young decapitated plant obtained in vitro from a sterile seedling; in most cases, the cotyledons are excised and the roots are trimmed.

The apex is placed on the decapitated surface of the rootstock in contact with the cambial zone. After grafting, the young plants are transferred to a liquid mineral nutrient which can guarantee the harmonious development of the graft (Navarro, Roistacher, and Murashige, 1975). The apex develops as a leafy shoot. The young plants obtained in this manner are then transferred to pots and acclimatized in a greenhouse before being planted out in an orchard. This technique is used for obtaining disease-free clones as well as for studies on incompatibility.

## Improvements Made to the Initial Technique

### Young Rootstock Plants

For many species, a stratification at low temperatures between 3 and 4°C is necessary to break the dormancy of the seeds. The duration of stratification is different for different species. For prune and apricot it is from 30 to 60 days, for peach 80 to 120 days, and for myrobalan (a variety of plum) 90 to 120 days. This stratification at 3°C in a cold chamber is carried out aseptically in darkness, either in culture tubes in an agar-solidified medium (one seed per tube) or in a Petri dish between two layers of damp filter paper. For certain species, it is possible to break the dormancy of the seeds by soaking them for 24 h in a cytokinin solution (e.g., BAP), in gibberellic acid (GA3), or a mixture of the two.

In peach, apricot, and cherry, the young plants obtained in this way without preliminary stratification are stronger and their germination more regular. In myrobalan the application of cytokinins does not eliminate the need for the cold stratification, but reduces the time by half.

### Apex Grafting

A technique which pretreats the apex, allowing the selection of the viable apex and helping their development, greatly improves the micrografting percentage of success. In peach, zeatin ($0.01$ mgL$^{-1}$) pretreatment for 2 to 10 days, or with 0.1 milligrams per liter for only 4 to 48 hours improves the percentage of success during the later grafting, which produces several long leafy green shoots. Perhaps this exogenous phytohormonal support allows the rebalancing of the endogenous hormone content of the graft, thus changing the physiological state of the excised organ.

## Micrografting

*Blockage of oxidation.* The first difficulty encountered is the oxidation of phenolics at the cut surfaces (at the level of the dissected apex and the dissected part of the stock). The contact of the phenolic compounds and their oxidation causes a discoloring of the tissues linked to micrografting. To block the oxidation phenomena, various substances, such as ascorbic acid, thiourea, cysteine, chlorohydrate, dithiothreitol (DTT), and sodium diethyl-dithiocarbamate (DIECA), have been used. The DIECA at a concentration of 2 grams per liter in the rinses and in a drop deposited on the tissues at the time of micrografting has been most effective.

*Feeding of the graft.* For certain graftings, a considerable improvement is obtained by inserting an agar block between the graft and the stock, which contains a mineral solution with or without added phytohormones. The agar prevents an intense and rapid dehydration of the apex, and induction of histogen (root meristem) on the grafting zone.

*Environment of the young* in vitro *grafted plant.* For the development of grafted explants, various supports have been tried, e.g., liquid medium with filter paper bridges (Heller 1953), and agar-solidified medium which has been detrimental to formation of roots. Vermiculite support has been more successful because it allows more oxygen to the rootstock system. Similarly, filter paper bridges or the Sorba Rod system (the Sorba Rod plugs consist of a cold-crimped cellulose paper wrapped with porous cellulose paper, on the model of a cigarette filter) provide good oxygen to the roots of rootstocks.

Sucrose content of up to 85 grams per liter has improved success of micrografting in citrus. This increases the percentage of success from 40 percent on the normal sucrose concentration of 30 to 50 grams per liter to 95 percent.

Light conditions play an important role. Thus, at the beginning of their development, the young grafted plants should be kept in darkness for three to seven days before being transferred to the light.

*The transfer.* Thirty to forty days after the in vitro grafting, the young grafted plants are transferred to a greenhouse, onto a suitable mix (sand, peat, vermiculite, 1:1:1), watered with dilute mineral solution (20 percent MS) and acclimatized gradually (Litz, Moore, and Srinivasan, 1985).

### Applications of Micrografting Techniques

1. Elimination of viruses: The in vitro micrografting technique has proved to be very useful in the regeneration of whole orchards of citrus fruits infected by viruses. This is because it is difficult to obtain rooting of the in vitro-isolated apex in trees. However, grafting the apex onto a suitable rootstock would eliminate this problem.
2. Grafted plants bypass the juvenile stage that is associated with sexual seedlings.
3. Production of plants resistant to pests, diseases, cold, water stress, waterlogging, salinity, etc.

## ANALYSIS OF COMPATIBILITY AND INCOMPATIBILITY PHENOMENA

### Biological Approach

There are two types of incompatibility:

1. Localized, characterized by a poor joining of the cambial formations of the two grafting elements, e.g., the graft between apricot and myrobalan.
2. Translocated incompatibilites are associated with an accumulation of starch around the grafting zone, with a normal vascular continuity in the grafted zone but degeneration of the phloem. Some associations of peach/myrobalan and peach/apricot are of this type. Incompatibility can be brought about between two species which are normally compatible by the presence of a virus or viroid at the graft.

This is particularly true for the association between citrus species and the bigarade (*Citrus aurantium* L.) in the presence of the tristeza virus or *Poncirus trifoliata* in the presence of the exocortis viroid.

In the case of compatible association, even between partners of the same species such as peach/peach, the success of grafting depends on the physiological "state" of the scion and the rootstock at the time of in vitro micrografting. For example, in peach the response was dependent upon the scion when the apex of the peach of one cultivar was grafted on young rootstock plants of the same cultivar. Success was best (83 percent) if the apex was taken in mid-spring to mid-summer, corresponding with maximum growth of shoots.

In young rootstock plants, the age of the plant also plays a part in successful grafting as is the case in the rootstock of the peach tree. Success is higher with young plants (5 to 7 days old: 61 percent) than with older ones (10 to 12 days old: 12 percent).

## Physiological Approach

Browning at the junction of in vitro micrografting of peach decreased the chances of success. This was due to the enzymatic oxidation of the phenolic compounds by the action of polyphenol oxidases and peroxidase. The quinones thus formed were subsequently polymerized into brown toxic products or inhibitors of several biochemical reactions.

## Cytological Approach

Various histological, histochemical, and histoenzymatic observations have been made on the compatible and incompatible association in the micrografts of prunus, and on the incompatibility created by the presence of a viroid, the exocortis, in the normally compatible association of Lisbon lemon and Troyer citrange (Huang, Che, and Chiu, 1986). Generally, for compatible combinations, the first signs appear 48 hours after the grafting on the rootstock and the graft, and all the fundamental tissues (cortical parenchyma, cambial cells) are likely to proliferate and participate in the positioning of the graft callus. About 15 days after in vitro micrografting, a cambium appears which guarantees a junction with the cambiums of the rootstock and scion (same as in conventional grafting).

As far as the incompatible combinations are concerned, the histo-physiological study of the junction zone allows us to see, in the case of the Canino apricot and myrobalan, a weak mitotic activity in both partners with a faulty vascular connection. It also shows an abnormal functioning of the newly formed cambium in the union zone, i.e., the presence of a "barrier" of necrotic cells on the rootstock/scion inter-face, as well as a peroxidase marking of persistent cellular debris.

In the same way, Lisbon lemon and Troyer citrange combinations infected with virus often show an almost nonexistent mitotic activity, while the healthy corresponding association shows an intense collagen at the junction zone. These observations enable us to detect, at an early stage (10 to 15 days after micrografting), an incompatibil-ity due to the viroid of the exocortis, which normally appears very late in the orchard, leading to an early detection of incompatible in-fected material.

### Problems Associated with Micrografting

Micrografting is a cumbersome, tedious, and extremely time-con-suming exercise. However, the benefits outweigh the problems asso-ciated with this technique, especially in tree crops where, at this point in time, the only way to get rid of diseases is micrografting. Further-more, this technique is useful in shortening the juvenile phase in tree crops and considerably reduces the amount of time in breeding programs.

Chapter 12

# In Vitro Flowering: Its Relevance to Plant Breeding

From the preceding chapters it is evident that aseptic techniques have been used extensively to enable us to understand various growth and developmental processes in plants. Unfortunately, the phenomenon of flowering, probably the most important of all developmental processes in angiosperms, has received much less attention than other domains in in vitro biology research. Nonetheless, considerable effort has been expended in the recent past to develop in vitro flowering systems for a number of species, mainly to unravel the mysteries of flowering (Bernier, 1988; Bernier et al., 1993). The in vitro approach has proven to be a very useful strategy for the investigation of flowering physiology (Van Staden and Dickens, 1991), and it is now emerging as a valuable approach that can be integrated into breeding programs for some species. In this chapter, various aspects of in vitro flowering and its significance in future crop improvement programs, particularly of difficult-to-breed species, are considered.

## *FACTORS INFLUENCING IN VITRO FLOWERING*

Flowering is a highly complex developmental process regulated by an array of biological and physical factors. These regulatory factors include photoperiod, light intensity, temperature, growth stage, endogenous level of hormones, (particularly gibberellins and cytokinins), and the available nitrogen form (Bernier et al., 1993). The involvement of these factors, initially identified in in vivo studies, has been proven to be critical for the induction of flowering in vitro too. In vitro flowering studies, however, unraveled many other regulatory elements involved in the flowering process. For instance, the regulatory role(s) of different nutrients and many bioactive compounds, such as benzoic acid derivatives and polyamines, was convincingly established by in vitro experiments.

## PLANT GROWTH REGULATORS

Among the various regulatory factors that induce in vitro flowering, plant hormones appear to play a key role in most of the species tested so far (Van Staden and Dickens, 1991). Auxin is widely considered to be an inhibitor of flowering in vitro, though reports show that it stimulates flowering. In day-neutral tobacco cultures, a small amount of auxin is required for flowering, but the bud development can be completely eliminated just by increasing the auxin concentration to supraoptimal levels. Application of α-naphthaleneacetic acid enhanced flowering in in vitro seedlings of *Pharbitis nil* and promoted flower bud formation in *Perilla fructescence.* Auxin-induced flowering was also reported in *Torenia* stem cultures and callus cultures of *Phlox drummondii.* Although these studies suggest a promotive role for auxin in in vitro flowering, examples where auxin acts as an inhibitor of flowering are more numerous. To cite a few examples, auxin strongly inhibited flowering in *Plumbago indica, Chrysanthenum, Kalanchoe blossfeldiana, Helianthus annus, Cichorium intybus, Begonia* spp., and *Streptocarpus nobilis.* In some day-neutral tobacco cultures, IAA caused a reversion of buds from reproductive to vegetative development. IAA was also found to inhibit cytokinin-induced flowering in a short-day plant, *Lemna acquinoctialis.* Interestingly, such auxin-induced inhibition could be negated by cytokinin application in many instances, indicating that, as with some other organogenic processes, the balance between auxin and cytokinin probably plays a regulatory role in flowering as well.

Though there are a few examples of auxin-stimulated flowering in vitro, it is unclear whether auxin is involved in the induction of flowering in vitro. In most of the culture conditions where auxin-stimulated flowering has been reported, explants were derived from a florally determined tissue, such as inflorescence. In other words, such explants had already acquired the developmental competence for reproductive development under permissive conditions. This indicates that the role of auxin in flowering may be limited to enhancing the differentiation and growth of flower buds rather than acting as a cue for the induction of flowering.

Of all the known plant hormones, gibberellins are the most intensively investigated hormones in relation to flowering (Bernier, Kinet, and Sachs, 1981a, b). This is not surprising as they have proven to be the most florigenic compounds tested thus far in in vivo experiments. Despite such a profound inductive role under in vivo conditions, gibberellins failed to evoke in vitro flowering in most of the species investigated. Gibberellic acid induced flowering in the shoot tips of *Chenopodium rubrum* seedlings, and in combination with BAP, supported flowering in somatic embryos of *Panax ginseng*. On the other hand, gibberellins inhibited flowering in several *Nicotiana* species, *Chrysanthemum* sp., *Pharbitis nil,* and *Torenia fournieri,* etc. It is believed that gibberellin application results in an unfavourable carbohydrate status for flowering in in vitro plants, but there is little direct evidence to support this hypothesis.

Cytokinins seem to be by far the most successful plant hormones in inducing flowering in vitro. Inclusion of cytokinins in the culture medium was necessary for flowering in many plant species including *Arabidopsis, Catharanthus tinctorius, Dionea muscipula, Helianthus annus, Passiflora suberosa, Silene cardinalis* and *Streptocarpus nobilis*. It is important to note that cytokinins were able to induce flowering in strictly vegetative explants obtained from plants maintained under noninductive conditions. Induction of flowering by cytokinin occurred only in a narrow range of concentrations, and supraoptimal levels of cytokinins in the medium invariably favored vegetative development in all the plants investigated so far.

Interestingly, different plant species require different types of cytokinins for the induction of flowering in vitro. For example, in the day-neutral plant *Passiflora,* as well as in bamboos, BAP is the most effective cytokinin for flower induction. In thin-layer explants of tobacco pedicels, both BAP and dihydrozeatin induced maximal bud formation at the same medium concentration. However, 2-isopentenyladenine was required at 20 to 40 times higher concentrations, while zeatin was noninductive in this system. In contrast, kinetin was more favorable than other cytokinins in flower production in *Arabidopsis* stem explants. For many species, nonetheless, a combination of different cytokinins or cytokinin and auxin was needed for in vitro induction of flowering.

Using *Sinapis alba* as the experimental system, Bernier and colleagues (Bernier et al., 1998) have studied the role of cytokinin in floral transition extensively. They found a significant transient increase in the levels of natural cytokinins, particularly zeatin, in response to the inductive signals. These observations, together with the results obtained from in vitro studies, indicate a regulatory role for cytokinin in floral induction (Bernier et al., 1998). However, we do not know the precise mechanism by which cytokinins regulate this complex process. It is hypothesized that cytokinins could be at least one of the inductive cues involved in the vegetative to reproductive phase change, but it is also argued that the primary role of cytokinins in flowering could be the initiation and control of early mitotic activity and associated cell synchronization for the development of the floral meristem. This latter contention gains support from the fact that cytokinin application often leads to meristem proliferation and it favors flowering only in a narrow concentration range.

Little work has been done on the action of inhibitors on in vitro flowering, despite their implication in the regulation of reproductive development. There are few examples in which exogenously applied abscisic acid (ABA) promotes flowering in vegetative explants kept under noninductive conditions. Presence of ABA in the medium initiated flowering in *Pharbitis* and *Torenia* cultures. The promotive effect of exogenously applied ABA in enhancing flower production was also evident in *Perilla* and *Kalanchoe*. Explants taken from young plants were found to be more responsive to ABA treatment than those from old plants.

Very little is known about the involvement of ethylene in in vitro flowering. Flowering was inhibited in *Chenopodium* and *Pharbitis* by the application of ethrel, an ethylene-releasing compound, and ethylene accumulation in airtight culture containers has been implicated in the failure of flowering in *Kalanchoe* explants maintained under inductive conditions. However, ethylene and its precursor methionine were able to induce flowering in *Plumbago* grown in vitro. Table 12.1 gives examples of species which have been successfully induced to flower in vitro.

In addition to the known phytohormones, numerous bioactive compounds have been found to induce or enhance flowering in vitro. Induction of flowering in *Lemna gibba* by salicylic acid and nicotinic acid has been convincingly demonstrated. Many derivatives of benzoic acid have been shown to be florigenic under in vitro conditions in a number of plant species.

TABLE 12.1. Species That Have Been Successfully Induced to Flower In Vitro

| Plant species | Explant | Parent plant induced | Growth regulators in the induction medium |
|---|---|---|---|
| *Allium sativum* | Floral tissues | Yes | GA$_3$ |
| *Begonia* species | Stem segment | No | Kinetin, adenine |
| *Bambusa arundinacea* | In vitro plantlets | No | BAP, coconut water |
| *Carthamus tinctorius* | Cotyledon | Not known | IAA, kinetin |
| *Cestrum diurnum* | Nodal segment | No | Absent |
| *Chrysanthemum* cultivars | shoot tip | No | GA$_3$ |
| *Cucumis sativus* | Hypocotyl | No | BAP, 2,4-D |
| *Dianthus caryophyllus* | Anther | Yes | Kinetin, coconut water, IAA |
| *Helianthus anuus* | Shoot apex | No | BAP or kinetin |
| *Kalanchoe blossfeldiana* | Floral tissues | Yes | BAP, NAA, 2,4-D |
| *Manihot esculenta* | Shoot apex | Not known | BAP, IAA, GA$_3$ |
| *Nicotiana tabacum* | Thin cell layers from inflorescence axis | Yes | BAP, kinetin, IAA |
| *Passiflora suberosa* | Leaf and stem segments | Yes | BAP |
| *Panax ginseng* | Somatic embryos | No | BAP, GA$_4$ |
| *Pisum sativum* | Shoot apex | Not known | BAP, NAA |
| *Silene cardinalis* | Shoot apex | Yes | Kinetin |
| *Spinacia oleracea* | Shoot apex | No | GA$_3$, GA$_7$ |
| *Streptocarpus nobilis* | Leaf segments | No | BAP |
| *Xanthium strumarium* | Shoot apex | No | Kinetin |

Note: BAP· 6-benzyladenine; 2,4-D: 2,4-dichlorophenoxyacetic acid; GA$_3$, GA$_4$, and GA$_7$: gibberellins 3, 4, and 7; IAA· indloe-3-acetic acid; NAA: ∝-naphthaleneacetic acid.

# MINERAL NUTRIENTS AND OTHER MEDIUM COMPONENTS

Intensive investigations on the Influence of various macro- and microelements on in vitro flowering are seldom made. Most of the in vitro flowering studies employed used Murashige and Skoog (1962) mineral salts (MS) as the basic nutrient medium. By manipulating the nutrient composition of a MS basal medium alone, flowering was induced in small vegetative explants of *Torenia* in the absence of hormones. From the extensive studies with *Perilla* and *Torenia,* it became apparent that a medium with low salts, particularly nitrogen in the form of ammonium, and high carbohydrate, promotes in vitro flowering. For some species, like *Torenia,* prolific flowering was observed when nitrates replaced all the ammonium salts in the medium.

An aspect that received considerable attention was the effect of carbohydrates on in vitro flowering. Several studies convincingly proved that carbohydrates play a significant role in the regulation of flowering, both in vivo and in vitro. Availability of soluble carbohydrates was essential for the induction of flowering in *Xanthium strumarium,* and glucose was able to mimic the effect of light on flowering in tobacco culture. High sucrose concentration induced flowering in *Sinapis alba* and substituted for the high light intensity requirement in *Cuscuta.*

Notably, different plant species required different types of sugars for in vitro induction of flowering. For example, glucose was necessary to elicit flowering in thin cell layer cultures of *Nicotiana,* while fructose caused more prolific flowering than glucose and sucrose in *Bougainvillea* cultures. A closer analysis of the effect of carbohydrates on in vitro flowering indicates that a high carbon to nitrogen ratio brings about flowering in many species.

## EXPLANT, LIGHT, AND OTHER VARIABLES

The choice of explant is a very important consideration for successful culmination of the desired morphogenic development and is influenced by various factors, including the objective of the investigation. It is imperative that explants are derived from a strictly vegetative plant maintained under noninductive condition if the goal is to unravel the regulatory aspects of the vegetative to reproductive phase transition. On the other hand, if the purpose is to develop an efficient in vitro flowering system which produces normal and fertile flowers that could be used for controlled breeding, an explant from a florally induced plant is preferable, because flowering can be realized relatively easily in such explants with minimal culture manipulations.

Experiments with *Nicotiana* and *Passiflora* indicate that the developmental determination to flower can be deeply imprinted in certain tissues and organs, such as inflorescence and floral tissues, and this developmental state of floral determination can be carried through from the flowering donor plant to the explant. These explants readily produced flowers when grown under permissive culture conditions. As with most other organogenic developments, young tissues were more responsive than older ones to in vitro treatments.

Light is required for successful induction of flowering in vitro. Apparently, no clear trend is emerging from the studies related to the effect of light quality and quantity on in vitro flowering. For example, in *Xanthium strumarium* high light intensity is needed for induction, whereas photoperiod is more critical for flowering in tobacco thin cell layer culture. In *Perilla,* photoperiod is crucial for flower development, but not critical for the transition from vegetative to reproductive growth.

In general, it is suggested that light is required for both induction of flowering and for normal flower development under in vitro conditions, but how light mediates flowering is far from clear.

## *APPLICATION OF IN VITRO FLOWERING TO PLANT BREEDING*

From the previous discussion it is conceivable that an in vitro system that is capable of producing fertile flowers relatively rapidly will have considerable practical importance in breeding plant species which are difficult to flower or produce flowers only once in several years. An excellent example illustrating this aspect is the in vitro flowering system developed for bamboos.

Bamboos flower only once during their lifetime, dying at the end of the fruiting season. This monocarpic flowering occurs after 12 to 24 years of growth, and is gregarious, i.e., all the bamboos in a local population flower simultaneously. Due to the unpredictability of flowering and seeding behavior, bamboo breeding has proved to be very difficult.

In 1990, however, this situation was changed when Nadgauda and colleagues reported an efficient in vitro flowering system for different genera and species of bamboo (Nadgauda, Prasharami, and Mascarenhas, 1990). In brief, in vitro-germinated bamboo seedlings developed normal flowers and seeds in a few months when cultured in liquid MS medium enriched with 2 percent sucrose, 5 percent coconut water and 0.5 milligrams per liter 6-benzyladenine and maintained at 28° C under 500 lux light on a gyratory shaker. More important, a proliferation of inflorescence with normal flowers was observed when explants were cultured on the same medium. This is undoubtedly an important advancement in bamboo breeding, as the in vitro flowering system can serve as a continuous source of fertile flowers capable of viable seed production. Notably, this method of bamboo flowering has been successfully tested with different genera and species of bamboos independently in different laboratories.

A similar example for continuous production of fertile flowers and viable seeds was recorded for orchids. Orchids normally take about three to six years to reach maturity. With in vitro flowering technique, flowers and viable seeds could be produced within six months of culture.

From the numerous reports on in vitro flowering, encompassing both monocotyledonous and dicotyledonous species, it is reasonable to assume that this methodology could be extended to virtually any flowering plant. However, considering the diversity of responses observed, and the range of manipulation required in different plant species, it remains difficult to define a generalized strategy for the induction of flowering in vitro. Nonetheless, there is sufficient evidence to suggest that the physiological status of the donor plant, low salt medium, and the presence of cytokinin appear to favor flowering and should therefore be considered when in vitro flowering is attempted with a new species. Now that the molecular details of the transition of vegetative meristem to floral meristem are rapidly being understood, it is becoming feasible to manipulate plants to flower at a desired growth stage, or in response to a particular environmental or endogenous signal (Weigel and Meyerowitz, 1994; Weigel and Nilsson, 1995; Colasanti and Sundaresan, 1996). Clearly, in vitro flowering systems capable of producing viable seeds would be a valuable approach to enhance the breeding programs of those species with long juvenile growth phase and unpredictable flowering behavior. The major challenge at the moment is to develop reliable in vitro methods that allow continuous production of fertile flowers and viable seeds of desired plant species.

# Chapter 13

# In Vitro Tuberization

## INTRODUCTION

Underground and aerial tubers, particularly potato *(Solanum tuberosum)*, yam *(Dioscorea alata)* and cassava *(Manihot escutenta)* are important crops cultivated extensively in different parts of the world. Unlike other major food crops, tuber crops are vegetatively propagated and as such are susceptible to many viral, bacterial, and fungal diseases. For example, more than 25 viral diseases have been reported in potato. This high incidence of pathogens increases the risks of spreading diseases and restricts international germplasm transfers for breeding programs. Pathogen-free plant materials are thus needed to facilitate international breeding programs as well as to contain diseases in commercial cultivation. This is usually accomplished by producing pathogen-free in vitro plants, either by meristem culture or by chemotherapy. In vitro plantlets, however, require specialized handling during transportation and transplantation, thus may not be economically viable for underdeveloped countries.

Interestingly, the realization that micropropagated tuber crop plants could be induced to produce microtubers in large numbers in vitro set a new direction for breeding and cultivation of tuber crops. The microtubers produced from pathogen-free plant material proved to be a very useful and efficient method to propagate, store, and transport at least potato and yam germplasm. In this chapter, we examine the experimental aspects of in vitro tuberization and discuss the significance of microtubers on breeding and commercial cultivation of potato and yam, two important tuber crops.

## FACTORS CONTROLLING MICROTUBER PRODUCTION

Since the early studies on in vitro tuberization in potato (Baker, 1953; Harmey, Rowley, and Clinch, 1966), production of microtubers has been reported in many *Solanum* and *Dioscorea* species and cultivars. For many years, both in vitro and in vivo studies have sought to identify the regulatory factors that control tuberization. It appears that a range of plant growth regulators, carbon source, photoperiod, nitrogen supply, culture temperature, medium getting agent, and type of explant and genotype influence initiation of in vitro tuberization.

Microtubers are normally produced from axillary meristems along the shoot (Hussey and Stacey, 1984). However, some impediments to the exploit of this technology are evident, especially in potatoes. Unlike many other members of Solanaceae, potatoes do not respond well to cytokinin stimulation of axillary shoot development and are thus difficult to multiply in culture. This step is critical, as it is the preexisting meristems that undergo tuberization upon induction. Again, this situation is further complicated by the strong competition between meristems along a stem axis, eventuating into the formation of only a single fully developed microtuber on each shoot. This problem is currently solved by manually preparing a single nodal segment for further shoot production or for induction of tuberization.

Tuberization in potato is usually evident in three to four weeks of culture under inductive conditions, while the first appearance of microtubers in yam *(D. alata)* may take about 8 to 12 weeks. Tuberization often occurs in senescent cultures, and it can be hastened by adding sucrose and cytokinin benzylaminopurine (BAP) to the medium. Murashige and Skoog (1962) (MS) medium is commonly used as the basal medium for microtuber production in both potato and yam. Enriching MS medium with high concentrations of sucrose, up to 6 percent, promotes tuberization in many potato cultivars. This phenomenon could be further augmented in some cultivars by the addition of cytokinin to tuber induction medium.

Photoperiod and light intensity seem to have a profound effect on microtuber production (Seabrook, Coleman, and Levy, 1993). They regulate various interactions of other tuber- inducing factors, including plant growth regulators. For example, the sucrose-induced in vitro tuberization in many potato cultivars occurred only under short-day conditions. Short days are reported to increase the endogenous cytokinin levels in potato cultures and the requirement of short days for in vitro tuberization could thus be negated, at least in part, by the application of cytokinin. Interestingly, in almost all the potato cultivars tested, the in vitro tuber-inducing ability of both sucrose and cytokinins was expressed maximally in cultures maintained in continuous darkness.

In contrast to potatoes, nodal cuttings of yam cultured on tuber-inducing medium produced the highest number of tubers under long days, or under continuous illumination, whereas larger tubers were obtained under short-day conditions (Jean and Cappadocia, 1991). In some cultivars of yam, MS medium was inhibitory to tuberization and this inhibition could be completely reversed by eliminating ammonium salts from the MS medium.

Interestingly, the amount of light influences tuber morphology and the site of tuberization in potatoes. Long days with bright light favored the production of long tubers in tuber-inducing medium. In the dark, long tubers were completely eliminated, although most of the cultures produced microtubers. Also significant is the fact that tubers can be produced at the end of the shoots simply by exposing the cultures to long-day conditions.

Cool temperatures are another critical factor involved in initiating in vitro tuberization. In potato, maximum tuber induction could be realized by maintaining the cultures between 15 and 25°C in an induction medium.

Microtubers of good size and substance, those with 50 mg or more in weight and more than 10 percent dry matter, can be stored for several months without losing viability. Microtubers of both potato and yam have dormancy periods equal to, or longer than, regular seed tubers. It is suggested that some of the culture conditions needed for in vitro tuberization cause prolonged dormancy in microtubers.

Nonuniform microtuber dormancy is a major problem limiting the utility of this technology for commercial application. The size, physiological maturity, and dormancy characteristics vary greatly between individual microtubers in a given population harvested at the end of a production cycle. Standardizing the tuberization time, sorting of harvested microtubers into different grades, or artificial breaking of dormancy, may solve the issues related to nonuniform microtuber production.

## PRACTICAL ASPECTS
## OF IN VITRO TUBERIZATION

As mentioned earlier, production of pathogen-free microtubers has considerable practical significance in relation to international breeding programs and germplasm exchange. Unfortunately, this valuable in vitro technique has not been applied to many of the commercially cultivated tuber crops. The knowledge base created with potato and yam, however, is sufficient to provide useful strategies to test this technology on other tuber crops. Nonetheless, it is prudent to suggest that the exact conditions of in vitro tuberization vary considerably with genotype, thus an optimum protocol for each clone should be empirically determined.

Currently, in vitro tuberization technology is not highly efficient or cost effective for commercial operation, though it is adequate for the needs of breeding programs. The major hurdles are (1) the provision of sufficient numbers of short meristems for tuber induction, and (2) the lack of control of uniformity in the time of tuberization. More effort toward the refinement of shoot proliferation and the development of sophisticated robotic systems for the preparation of single nodal cuttings would considerably enhance the efficiency of the present system. However, the development of a bioreactor that can combine the requirements of microtuber quantity and quality in a cost-effective manner will ultimately aid in the realization of the full potential of microtuber technology.

# Chapter 14

# Molecular Plant Breeding

"Molecular breeding" is a general term used to describe the development and application of molecular genetic techniques to introduce novel, desirable characteristics with high value to plant breeding programs (Karp, Isaac, and Ingram, 1998).

Selection and breeding of plants was a passive process prior to the nineteenth century. It was primarily based on the natural selection events. At the end of the nineteenth century, deliberate crossing of plants to introduce desirable traits began. With the rediscovery of Mendelian genetics in the early 1900s, active programs in plant breeding gained importance. Most of these Mendelian exercises were long, drawn out projects involving up to twelve rounds of crossing and selection that spanned several years prior to releasing improved crop varieties and hybrids. In the 1960s, the full potential of hybridization exercises was realized with the release of several high-yielding cereal varieties, which were the basis of the now famous "green revolution." Undoubtedly, this has contributed to alleviating the world food shortage.

The basic principle behind plant breeding is the introduction of desirable genes into selected crop species from distant relatives or wild species. Although not as direct as the modern genetic engineering techniques, this form of selective breeding amounts to genetic manipulation of crop plants. In the 1980s, methods for *Agrobacterium*-mediated gene transfer to plants and recovering transgenic plants was perfected. Isolating agronomically important genes and transforming cultivars with the new genes generated improved varieties of transgenic plants (popularly referred to as genetically modified organisms or GMOs), which are the focus of current public debate. However, we will not be discussing this aspect of plant improvement in this chapter.

Molecular marker technology is revolutionizing plant breeding.

The source of genes for desirable traits using genes for plant improvement are no longer limitations. With the modern tools, genes can be introduced into selected crop plants from practically any organism (or even synthetic genes for that matter!). Combined with the molecular biology techniques of gene cloning and the more recent "functional genomics," approaches for identifying whole arrays of genes that regulate desirable traits (such as resistance to disease and insect pests) have given a whole new dimension to plant breeding. Thus, molecular plant breeding has come into existence.

## TYPES OF MOLECULAR MARKERS

A variety of molecular marker systems have been developed to facilitate the analysis of complex plant genomes (Karp et al., 1998) Some types of molecular markers used for breeding are:

- Isozymes (these harness the differences among the different members of multigene families coding for specific enzymes)
- Restriction fragment length polymorphisms (RFLP)
- Random amplified polymorphic DNA (RAPD)
- Amplified fragment length polymorphisms (AFLP)
- Microsatellites or simple sequence repeats (SSR)—also known as short tandem repeats (STR)

The term *polymorphism* describes the existence of different forms within a population, e.g., a difference in the number of tandem repeats in the microsatellites.

Isozymes or isoenzymes, by definition, are any of multiple forms of a given enzyme, each with slightly different kinetics, but with the same catalytic function. The different isozymes are formed by different members of a multigene family (which have minor differences in their sequences). It is often possible to correlate specific isozyme profiles with particular traits, which makes them useful markers for breeding.

Among the above markers, RFLP has been the most commonly used and informative molecular marker in mapping crop plant genomes. RFLP analysis utilizes the differences in the nucleotide sequences at which restriction enzymes cut the genomic DNA. The technique involves extraction of genomic DNA, digesting the DNA with selected restriction enzymes and probing the blot with known DNA segments (Southern blots). Its use has been limited by the relatively large amount of DNA required for the analysis. Also, this technique is time-consuming and expensive, making it less suitable for large-scale screening programs in plant breeding.

Advances in functional genomics and polymerase chain reaction (PCR) have resulted in several other equally powerful techniques that can be used for the evaluation of a large number of plants (e.g., for germplasm screening), and for the identification of markers for use in breeding programs. PCR-based markers are simpler to use for large numbers of samples. Microsatellites (or SSR), RAPD, and AFLP markers come under this category. The cost of generating such markers is high; however, with the advances in technology and with international collaborations, the cost may be reduced and a greater number of such markers may become available to the breeders.

RAPD utilizes the differences in the segments of genomic DNA that will be amplified by PCR using short (usually 10 nucleotide-long) oligonucleotide primers. RAPD markers generated are usable as dominant markers. The use of such markers for specific traits has been reported in several cases. One of the disadvantages of RAPD is that the amplification is carried out under low stringency, making the resultant bands less reproducible in different laboratories. However, once a marker is identified, the DNA band can be cloned and sequenced to generate a robust marker for use by all. Among the PCR-based markers, this is perhaps the cheapest and technically simplest. The availability of prior genetic information is also not a prerequisite for the use of RAPD markers. Hence, it can be applied to new plant species with relative ease.

There are several reports confirming that AFLP can make a significant contribution to the study of plant genetics. This technique utilizes the differences in restriction sites on the genomic DNA and selective PCR amplification. This is arguably more reliable than the uncharacterized RAPD markers. With the improved technology currently available, AFLPs have become highly reliable and offer ample polymorphism (Paul et al., 1997; Saghai et al., 1994). A large number of AFLP loci that are usable as markers can be generated for a given plant from a single experiment.

Microsatellites, or SSR, consist of tandem repeats of a simple sequence of nucleotides, each between one and six base pairs in length.

The name 'satellites' comes from the optical spectra generated after density gradient separation of DNA fragments with significantly different base composition, when the bulk of the DNA appears as a main band, while bands corresponding to repeated DNA appear as 'satellite' bands.

Microsatellites are widely dispersed throughout eukaryotic genomes (interspersed in the coding regions of the genome, but they are untranslated stretches of DNA) and are often highly polymorphic due to variation in the number of repeat units. The repeats may be in the form of CT (or AT or CA etc.), GTT, or AAAAT, and the number of times these stretches of nucleotides are repeated can vary from a few occurrences to a few dozen. GT and GTT appear to be the most common repeat sequences in barley. The flanking regions of these repeats tend to be highly conserved, thus making it possible for us to synthesize PCR primers for the borders. The total size of a microsatellite locus is small and the repeats tend to be less than 150 bp in total length. Microsatellites mutate by gaining or losing repeat units at a high rate (on the order of $10^{-3}$ to $10^{-4}$ per locus per generation for dinucleotides). They are not counter selected, so microsatellites accumulate in a given species. Furthermore, SSR markers require only small amounts of sample DNA, and because of their hypervariability, codominance, and ubiquity in eukaryotic genomes, these markers are used for population genetic studies, strain identification, paternity analysis, and mapping purposes.

The choice of the marker system for any application will depend on the type of breeding program and it should be a high throughput, simple yet robust system.

## MAJOR OBJECTIVES OF MOLECULAR BREEDING

- Making available molecular markers for desirable characteristics (e.g., disease resistance, stress tolerance, high yield, uptake and efficient use of micronutrients etc.) for crop improvement programs
- To develop tools for identification and estimation of the quantity of pathogen strains in plants and soil
- Developmental enhancement of the crop plants through modifications to specific pathways of development or biosynthesis of specific chemicals

- To develop molecular markers for quality of the produce (e.g., flour color, starch and protein compositions in grains)
- To help in improving the speed and reliability of bioassays for measuring crop performance (e.g., based on nutrient deficiency, extent of disease etc.), including for marker-assisted backcrossing

## APPLICATIONS OF MOLECULAR MARKERS IN PLANT BREEDING

To date, the applications of molecular markers for cereal breeding include accelerated backcrossing, screening double haploid donor plants and progeny, screening of populations for elite lines, confirming the genotypes of parent lines, pyramiding of multiple resistance loci, and analysis for Plant Breeders' Rights applications. Genetic maps showing the relative positions of genetic markers, and individual genes affecting desirable traits (e.g., disease resistance, plant morphology, biochemical traits) are being developed. Quantitative trait loci (QTL; i.e., genes affecting quantitative traits) can also be mapped relative to molecular markers.

Quantitative traits are those that are present to a greater or lesser extent in all members of a species. These are usually controlled by multiple genes, and the traits have an average value, around which individuals in a population are distributed (e.g., plant height, number of grains per ear, fruit size etc.). Also, quantitative traits are subject to environmental influence. In contrast, the perceivable qualitative traits are those that are regulated by a single gene (or two genes at the most) and may be due to mutations in the gene. Consequently, some members of a species or population may not exhibit the quality under observation (e.g., disease resistance).

Marker-assisted selection, accelerated backcrossing, genetic fingerprinting, and estimating genetic distance are some of the most important applications of molecular markers in plant breeding.

### Marker-Assisted Selection

Breeders can select for marker genes that are proven to be linked to the genes controlling the desirable trait, instead of selecting for the trait phenotype. Linked genes are always inherited together and the presence of one (whose sequence is known) will imply that the gene for the desirable trait is also present in the individual. DNA markers serve as powerful selection tools for agronomically important genes in a breeding program allowing trait screening at the early stages of plant development. This is possible because such selection can be made without having to grow the plants to maturity during the early generation screens, thereby speeding up the entire process. However, the final selected generation of plants will have to be screened by growing them to maturity.

The genetic linkage maps used in rice, wheat, barley, and other cereal improvement programs so far have been primarily RFLP marker-based. However, the relatively low level of polymorphism detected in some cultivars limits the usefulness of RFLP. The advent of PCR-based marker systems, such as AFLP and SSR, has facilitated the identification of a larger number of DNA polymorphisms to construct more complete linkage maps in a much shorter time than was possible with RFLP alone. This offers the benefits of the known genetic/chromosomal location of the RFLP markers and the much larger number of AFLP/SSR markers generated at a faster pace.

It is now possible to identify molecular markers linked to genes controlling several major aspects of quality and QTL, which in turn can be used for improving the crop plants. Thus, QTLs controlling various aspects such as germination, flour color, milling yield, and milling energy requirement characters have been identified in several cereal species, including barley and wheat. QTLs conferring adult plant resistance to stripe rust (caused by *Puccinia striiformis* f.sp. *hordei*) in barley were mapped (Toojinda et al., 1998). The resistance QTLs were introgressed into a genetic background unrelated to the mapping population with one cycle of marker-assisted backcrossing. Doubled-haploid lines were derived from selected backcross lines, phenotyped for stripe-rust resistance, and genotyped with an array of molecular markers. The resistance QTLs that were introgressed were significant determinants of resistance in the new genetic background, demonstrating the usefulness of the techniques employed.

## *Accelerated Backcrossing*

The aim of backcrossing is to allow the introgression of one specific gene into an elite cultivar. This task takes a long time because the amount of recurrent parent genome must be maximized at each backcross. The chromosome carrying the desirable gene that is to be introgressed takes more time and generations to come back to the recurrent form (as a result of crossing over and chromatid exchanges). The availability of molecular markers helps to choose the individuals having the highest percentage of the recurrent parent at each generation, at the same time possessing the gene for introgression. This can facilitate selection of the ideal genotypes (containing 97 to 99 percent of recurrent parent genome + the integration) in two to three generations, e.g., in barley, wheat, maize, sunflower, and rapeseed. Clearly, such accelerated selection would translate into significant time and cost savings.

The use of molecular markers for accelerated backcrossing has been applied in barley breeding (Ellis et al., 2000). By using polymorphic AFLP markers, breeders found that the percentage of donor DNA composition varied from 8 percent to 60 percent in the backcross progeny being screened. They estimated that selection of the genotype with 8 percent donor parent genome would be equivalent to saving two generations of backcrossing. A similar strategy is being applied in the breeding of other cereals, including wheat. The widespread leaf rust and stem rust diseases, caused by the fungus *Puccinia graminis* f. sp. *tritici*, led to significant yield losses in susceptible wheat cultivars. The wild relative of wheat, *Aegilops ventricosa*, possesses a gene complex *(Sr38, Lr37, Yr17)*, which can confer resistance to the disease. Efforts are underway to introduce this and other gene complexes to commercial wheat cultivars by introgression and the aid of molecular markers for progeny selection (Cox, 1998).

## Genotyping and Genetic Fingerprinting of Plants

The use of molecular markers for genetic fingerprinting in general is now well established. Molecular markers are particularly useful in revealing genetic variation among closely related plants (e.g., cultivar level) where phenotypic markers fail to show variability. Examples for the application of markers such as RAPD polymorphisms for cultivar identification in ornamental species such as *Ixora* and *Licuala* have also been reported (Rajaseger et al., 1997; Loo et al., 1999).

As indicated earlier, in plant breeding, molecular markers can be used to distinguish cultivars which will be useful in selecting the parental lines for a specific cross to be set up from others. Determining the level of genetic diversity within a species, genetic relationships among different cultivars, landraces (genetically diverse, cultivated plant populations) and wild species to be used for breeding is critical. Such information is useful to optimize genetic variances in a breeding program.

In barley, SSR markers for several traits of interest to malting, fermentation, and distilling have been mapped for several (random inbred lines) contemporary cultivars (Swanston et al., 1999). This and similar findings would permit the use of molecular markers in various breeding programs. Similarly, using several previously mapped RFLP markers as anchor probes, an extensive genetic linkage map of lentil *(Lens culinaris)* was constructed (Eujayl et al., 1998) with 177 markers (89 RAPD, 79 AFLP, 6 RFLP and three morphological markers). Such maps could be used for the identification of markers linked to QTLs in the population of inbred lines used.

### Estimating Genetic Distance

DNA polymorphisms reflect the relationships among cultivars, species, and populations of plants. Information from molecular markers can help plant breeders choose the appropriate parental germplasm for carrying out crosses. AFLP markers were successfully employed to detect diversity and genetic differentiation among Indian and Kenyan populations of tea *(Camellia sinensis)* (Paul et al., 1997). This analysis grouped Assam genotypes, both from India and Kenya, confirming that the Kenyan clones have been derived from collections made in the Assam region. The China types came under a separate group, which is a reflection of wider genetic variation. Also, clones collected from the same region exhibited less overall genetic variation. AFLP analysis discriminated all of the tested genotypes, even those that could not be distinguished on the basis of morphological traits.

### CASE STUDY:
### APPLICATION OF MOLECULAR MARKERS
### IN BARLEY (HORDEUM VULGARE) BREEDING

Molecular markers are now routinely being adopted in cereal breeding programs. Barley is an ideal crop for applying molecular breeding technology, because it is a diploid (2n = 14) species. It has a short life cycle and the cultivars are inbred lines, making the genetic and physiological analyses relatively easy to perform (Forster et al.,

1999). Several source species, e.g., specifically adapted landraces, and wild barley *(H. spontaneum)*, are available to introgress desirable genes into the inbred cultivars. Further, the wild barley and landraces are all diploid and intercrossable.

It should be mentioned here that barley is one of the oldest crop species known (from archeological data) to have been in cultivation as early as 8000 B.C. The long history of selection has made the modern cultivars genetically rather uniform. However, several desirable traits exist in wild barley that can be introgressed into the cultivars. The traditional approach has been to try and introduce one trait at a time. With the availability of the molecular markers, and an effective procedure for obtaining doubled haploid lines from anther culture from barley, breeders are now aiming to introgress multiple traits simultaneously (e.g., drought tolerance and mildew resistance; cereal cyst nematode (CCN) resistance; boron tolerance (BT); resistance to barley yellow dwarf virus (BYDV) and powdery mildew *(Mlo);* manganese efficiency (Mn); and the Denso dwarfing gene).

With the demonstrated success of molecular breeding techniques, breeders are able to screen a larger proportion of their breeding material. However, the number of existing markers is insufficient. With international collaborative efforts and easy sharing of data, coupled with the advances in PCR-based marker techniques, a considerable database is likely to become available to breeders. With marker-assisted selection, fewer individuals need to be screened within a given population while enabling higher gene frequencies to be obtained, which greatly improves the efficiency of the breeding program. The use of molecular markers for accelerated backcrossing is now widely accepted, which facilitates early identification and removal of the undesirable lines, thus reducing the workload.

The earlier attempts at introgressing single traits to cultivars of barley from the wild species were hampered by the fact that several undesirable traits were cointegrated with the desirable traits (Saghai Maroof et al., 1994). However, the current information suggests that by using appropriate genetic backgrounds, some of the undesirable linkages may be broken, and this will permit generating breeding lines that were not possible without the modern tools. Considering the tremendous efforts being put into such breeding programs by a multitude of plant breeders, supported by an impressive array of molecular tools that are constantly being improved, it is not inconceivable that highly desirable cereal crops will be generated in the near future.

# References

## Chapter 1

Belliard, G., Vedel, F., and Pelletier, G. (1979). Mitochondrial recombination in cytoplasmic hybrids of *Nicotiana tabacum* by protoplast fusion. *Nature* 281: 401-402.

Bourgin, J.P. and Nitsch, J.P. (1967). Obtention de *Nicotiana* haploides à partir d'étamines cultivées in vitro. *Annales des Physiologie Vegetale* 9: 377-382.

Cocking, E.C. (1960). A method for the isolation of plant protoplast and vacuoles. *Nature* 187: 927-929.

Gautheret, R.J. (1934). Culture du tissus cambial. *Comptes Rendus des Seances Academie des Sciences* 198: 2195-2196.

Gleba, Y.Y. and Sytnik, K.M. (1984). *Protoplast Fusion: Genetic Engineering of Higher Plants*. Heidelberg: Springer-Verlag.

Guha, S. and Maheshwari, S.C. (1964). In vitro production of embryos from anthers of *Datura*. *Nature* 204: 497.

Guha, S. and Maheshwari, S.C. (1966). Cell division and differentiation of embryos in the pollen grains of *Datura* in vitro. *Nature* 212: 97-98.

Morel, G. (1960). Producing virus-free cymbidium. *American Orchid Society Bulletin* 29: 495-497.

Murashige, T. and Skoog, F. (1962). A revised medium for rapid growth and bioassays with tobacco tissue cultures. *Physiologia Plantarum* 15: 473-494.

Nobecourt, P. (1937). Cultures en série de tissus végétaux sur milieu artificiel. *Comptes Rendus Hebdomadaires des Seances Academie des Sciences.* 200: 521-523.

Nobecourt. P. (1939). Sur la perennite et l'augmentation de volume des cultures de tissus vegetaux. *Comptes Rendus des Seances. Societe de Biologie et de ses Filiales.* 130: 1270-1271.

Reinert, J. (1958). Morphogeneses und ihre kontrolle an gewebe kulturen aus karotten. *Naturwiss* 45: 344-345.

Schell, J. (1987). Transgenic plants as a tool to study the molecular organization of plant genes. *Science* 237: 1176-1183.

Schell, J. and Vasil, K. (Eds.) (1989). *Cell Culture and Somatic Cell Genetics of Plants*. Vol. 6., Molecular Biology of Plant Nuclear Genes. New York: Academic Press.

Skoog, F. and Miller, C.O. (1957). Chemical regulation of growth and organ formation. *Symposium of the Society for Experimental Biology* 11: 118-131.

Steward, F.C., Mapes, M.O., and Mears, K. (1958). Growth and organized development of cultured cells. *American Journal of Botany* 45: 693-713.
White, P.R. (1934). Potentially unlimited growth of excised tomato tips in liquid medium. *Plant Physiology* 9: 585-600.
White, P.R. (1943). *A Handbook of Plant Tissue Culture.* Lancaster, PA: J.Casttell Press.

## Chapter 2

Miller, C.O., Skoog, F., Von Saltzo, M.N., and Strong, F.M. (1955). Kinetin, a cell division factor from deoxyribonucleic acid. *Journal of American Chemical Society* 77: 1392.
Skoog, F. and Miller, C.O. (1957). Chemical regulation of growth and organ formation in plant tissues cultured in vitro. *Symposium of the Society for Experimental Biology.* 11: 118-131.
Taiz, L. and Zeiger, E. (1991). *Plant Physiology.* Sydney: The Benjamin/Cummings Publishing Company, Inc.
Tran Thanh Van, M. (1980). Control of morphogenesis by inherent and exogenously applied factors in thin cell layers. In *International Review of Cytology,* pp. 175-194. Supplement IIA, New York: Academic Press Inc.

## Chapter 3

Broertjes, C. and Van Harten, A.M. (1978). *Application of Mutation Breeding Methods in the Improvement of Vegetatively Propagated Crops.* Amsterdam: Elsevier Scientific Publishing Company.
Cresswell, R.J. and Nitsch, C. (1975). Organ culture of *Eucalyptus grandis. Planta* 125: 87-90.
Debergh, P.C. and Maene, L.J. (1981). A scheme for commercial propagation of ornamental plants by tissue culture. *Scientia Horticulturae* 14:335-345.
de Fossard, R.A. (1976). *Tissue Culture for Plant Propagators.* Australia: University of New England, Armidale.
Dumet, D. (1994). Cryoconservation des massifs d'embryons somatiques de palmier a huile (*Elaeis guineensis* Jacq.) par deshydratation-vitrification. Etude du role du saccharose pendant le pretraitment. PhD thesis, L'Universite P. et M. Curie, Paris.
Jones, W.N. (1937). Chimaeras: A summary and some special aspects. *Botanical Review* 3: 545-562.
Kartha, K.K. (Ed.) (1985). *Cryopreservation of Plant Cells and Organs.* Boca Raton, FL: CRC Press.
Kartha, K.K. and Engelmann, F. (1994). Cryopreservation and germplasm storage. In *Plant Cell and Tissue Culture,* Vasil, I.K. and Thorpe, T.A., (Eds.), (pp.195-230) Dordrecht: Kluwer Academic Publishers.

Kumar, P.P., Reid, D.M., and Thorpe, T.A. (1987). The role of ethylene and carbon dioxide in differentiation of shoot-buds in excised cotyledons of *Pinus radiata* in vitro. *Physiologia Plantarum* 69: 244-252.

Maene, L.J. and Debergh, P.C. (1986). Optimisation of plant micropropagation. *Medical Facility Landbouww Gent* 51: 1479-1488.

Read, P.E. (1988). Stock plants influence micropropagation success. *Acta Horticlturae* 226:41-52.

Schafer-Menuhr, A., Schumacher, H-M, and Mix-Wagner, G. (1994). Langzeitlagerung alter Kartoffelsorten durch Kryokonservierung der Leristeme in flussigem Stickstoff. *Landbauforschung Volkenrode* 44: 301-313.

Smith, R.H. (1992). *Plant Tissue Culture: Techniques and Experiments.* San Diego: Academic Press, Inc.

Withers, L.A. (1995). Collecting for genetic resources conservation. In *Collecting Plant Genetic Diversity: Technical Guidelines,* Guarino. L., Ramanatha Rae, Reid, R., (Eds.) (pp. 115-123). New York: CAB International.

Withers, LA. and Engelmann, F. (1995). In vitro conservation of plant genetic resources. In *Biotechnology in Agriculture,* Altman, A., (Ed.) (pp. 57-88). New York: Marcel Dekker Inc.

## Chapter 4

Bhojwani, S.S. and Razdan, M.K. (1983). *Plant Tissue Culture: Theory and Methods.* New York: Elsevier Publishing Co.

Bourgin, J.P. and Nitsch, J.P. (1967). Obtention de *Nicotiana* haploides à partir d'étamines cultivées in vitro. *Annales des Physiologie Vegetale* 9: 377-38.

Dunwell, J.M. (1976). A comparative study of environmental and developmental factors which influence embryo induction and growth in cultured anthers of *Nicotiana tabacum. Environmental and Experimental Botany,* 16: 109-118.

Guha, S. and Maheshwari. S.C. (1964). In vitro production of embryos from anthers of *Datura. Nature* 204: 497.

Prakash, N. (2000). *Methods in Plant Microtechnique* (Third edition). Armidale, Australia: University of New England.

Sunderland, N. and Wicks, F.M. (1971). Embryoid formation in pollen grains of *Nicotiana tabacum. Journal of Experimental Botany* 22: 213-226.

Thanh-Tuyen, N.T. and de Guzman, E.V. (1983). Formation of pollen embryos in cultured anthers of coconut (*Cocos nucifera* L.). *Plant Science Letters* 29: 81-88.

## Chapter 5

Balatková, V. and Tupy, J. (1968). Test-tube fertilisation in *Nicotiana tabacum* by means of an artificial pollen tube culture. *Biologia Plantarum* 10: 266-270.

Kanta, K., Rangaswamy, N.S., and Maheshwari, P. (1962). Test-tube fertilisation in a flowering plant. *Nature* 194: 1214-1217.

Pierik, R.L.M. (1987). *In Vitro Culture of Higher Plants.* Dordrecht: Kluwer Academic Publishers.

Taji, A. M. and Williams, R.R. (1987). Perpetuation of the self-incompatible rare species of *Swainsona laxa* R. Br. by pollination in vitro and in situ. *Plant Science* 48: 137-140.

Van Overbeek, J. (1942). Hormonal control of embryo and seedling. *Cold Spring Harbor Symposium and Quantitative Biology* 10: 126-133.

Zenkteler, M. (1970). Test-tube fertilisation of ovules in *Malandrium album* Mill. with pollen grains of several species of Caryophyllaceae family. *Experientia* 23:775-776.

## Chapter 6

Carlson, P.S., Smith, H.H., and Dearing, R.D. (1972). Parasexual interspecific plant hybridization. *Proceedings of National Academy of Science, USA* 69: 2292-2294.

Cocking, E.C. (1960). A method for the isolation of plant protoplast and vacuoles. *Nature* 187: 927-929.

Dodds, J.H. and Roberts, L.W. (1985). *Experiments in Plant Tissue Culture* New York: Cambridge University Press.

Stafford, A. and Warren, G (1991). *Plant Cell and Tissue Culture* New York: John Wiley & Sons Limited.

## Chapter 7

Carlson, P. S. (1973). Methionine-sulfoximine-resistant mutants of tobacco. *Science* 180: 1366-1368.

Dix, P. J. (1994). Isolation and characterization of mutant cell lines. In *Plant Cell and Tissue Culture* Vasil, I.K. and Thorpe, T.A. (Eds.) (pp. 119-138). Dordrecht: Kluwer Academic Publishers.

Jayasankar, S. (2000). Variation in tissue culture. In *Plant Tissue Culture Concepts and Laboratory Exercises* (Second edition) Trigiano, R.N. and Gray, D.J. (Eds.) (pp. 387-395). Boca Raton, FL: CRC Press.

Larkin, P. J. and Scowcroft, W. R. (1981). Somaclonal variation—a novel source of variability from cell cultures for plant improvement. *Theoretical and Applied Genetics* 60: 197-214.

Maliga, P., Breznovitis, A., and Marton, L. (1973). Streptomycin-resistant plants from callus cultures of haploid tobacco. *Nature: New Biology* 244: 29-30.

Schaeffer, G. W. and Sharpe F. T. (1990). Modification of amino acid composition of endosperm proteins from in-vitro-selected high lysine mutants in rice. *Theoretical and Applied Genetics* 80: 841-846.

Toyoda, H., Chatani K., Matsuda Y., and Obuchi S. (1989). Multiplication of tobacco mosaic virus in tobacco callus cultures and in vitro selection for viral disease resistance. *Plant Cell Reports* 8: 433-436.

# Chapter 8

Coen, E. S., Romero, J. M., Doyle, S., Elliot, R., Murphy, G., and Carpenter, C. (1990). *'Floricaula':* A homeotic gene required for flower development in *Antirrhinum majus. Cell* 63: 1311-1322.

Dix, P. J. (1999). Mutagenesis and the selection of resistant mutants. In *Methods in Molecular Biology, Volume 3, Plant Cell Culture Protocolsm,* Hall, R. D. (Ed.). (pp. 309-318). Totowa, NJ: Humana Press Inc.

Donnini, P. and Sonnino, A. (1998). Induced mutation in plant breeding: Current status and future outlook. In *Somaclonal Variation and Induced Mutations in Crop Improvement,* Jain, S. M., Brar, D. S., and Ahloowalia, B. S. (Eds.). (pp. 255-291). Dordrecht: Kluwer Academic Publishers.

Liljegren, S. J., Ditta, G. S., Eshed, H. Y., Savidge, B., Bowman, J. L., and Yanofsky, M. F. (2000). *SHATTERPROOF* MADS-box genes control seed dispersal in *Arabidopsis. Nature* 404: 766-770.

Maluszynski, M., Nichterlein, K., van Zanten, L., and Ahloowalia, B.S. (2000). Officially released mutant varieties—the Food and Agriculture Organization (FAO)/International Atomic Energy Agency (IAEA) Database. *Mutation Breeding Review* 12: 1-84.

McClintock, B., (1951). Chromosome organization and gene expression. *Cold Spring Harbour Symposium of Quantitative Biology* 16: 13-57.

Negrutiu, I. (1990). In vitro mutagenesis. In *Plant Cell Line Selection: Procedures and Applications,* Dix, P. J. (Ed.). (pp. 19-38). Weinheim, Germany: VCH Publishers.

Rodriguez, F. I., Esch, J. J., Hall, A. E., Binder, B. M., Schaller, G. E., and Bleecker, A. B. (1999). A copper cofactor for the ethylene receptor ETR1 from *Arabidopsis. Science* 283: 996-998.

Sundaresan, V., Springer, P., Volpe, T., Haward, S., Jones, J. D. G., Dean, C., Ma, H., and Martienssen, R. (1995). Patterns of gene-action in plant development revealed by enhancer trap and gene trap transposable elements. *Genes and Development* 9: 1797-1810.

Yanofsky, M. F., Ma H., Bowman, J. L., Drews, J. N., Feldmann, K. A., and Meyerowitz, E. M. (1990). The protein encoded by the *Arabidopsis* homeotic gene *Agamous* resembles transcription factors. *Nature* 346: 35-39.

# Chapter 9

Bhojwani, S.S. (1990). *Plant Tissue Culture: Applications and Limitations.* Amsterdam: Elsevier.

Gautheret, R.J. (1985). History of plant tissue and cell culture: A personal account. *In Cell Culture and Somatic Cell Genetics of Plants* Volume 2, Vasil, I.K. (Ed.). (pp. 1-59). New York: Academic Press.

Groose, R.W. and Bingham E.T. (1986). An unstable anthocyanin mutation recovered from tissue culture of alfalfa *(Medicago sativa)*. 1. High frequency of reversion upon reculture. *Plant Cell Report* 5:104-107.

Henry, R.J. (1997). *Practical Applications of Plant Molecular Biology*. London: Chapman and Hall.

Kumar, A. and Bennetzen, J.F. (1999). Plant retrotransposons. *Annual Review of Genetics* 33: 479-532.

Larkin, P.J. and Scowcroft, W.R. (1981). Somaclonal variation—A novel source of variability from cell cultures of plant improvement. *Theoretical and Applied Genetics* 60: 197-214.

McClintock, B. (1984). The significance of responses of the genome to challenge. *Science* 226: 792-801.

Prakash, A.P. and Kumar, P.P. (1997). Inhibition of shoot induction by 5-azacytidine and 5-aza-2'-deoxycytidine in *Petunia* involves DNA hypomethylation. *Plant Cell Report* 16: 719-724.

## Chapter 10

Bajaj, Y.P.S. (1995). *Cryopreservation of Plant Germplasm: Biotechnology in Agriculture and Forestry, volume 32,* Berlin: Springer-Verlag.

Dodds, J.H. (1991). *In Vitro Methods for Conservation of Plant Genetic Resources*. London: Chapman and Hall.

Guy, C.L., Niemi, K.J., Fennel, A., and Carter, J.V. (1986). Survival of *Cornus sericea* L. stem cortical cells following immersion in liquid helium. *Plant Cell and Environment* 9: 447-450.

Kartha, K.K. (1985). *Cryopreservation of Plant Cells and Organs*. Boca Raton, FL: CRC Press.

Kartha, K.K., Fowke, L.C., Leung, N.L., Caswell, K.L., and Hakman, I. (1988). Induction of somatic embryos and plantlets from cryopreserved cell cultures of white spruce *(Picea glauca)*. *Journal of Plant Physiology* 132: 529-539.

Kumu, Y., Harada, T., and Yakuwa, T. (1983). Development of a whole plant from a shoot tip of *Asparagus officinalis* L. Frozen down to –196°C. *Journal of Faculty of Agriculture.* (Hokkaido University, Japan) 61: 285-294.

## Chapter 11

Heller, R. (1953). Recherches sur la nutrition minérale des tissus végétaux cultivés in vitro. *Annales des Sciences Naturelles. Botanique et Biolgie Vegetale* 14: 1-223. (Not seen in original)

Huang, Li-Chun, Che, Wen-Lin, and Chiu, Dor-Shen (1986). In vitro graft-enhanced nucellar plant development in the mono embryonic *Citrus grandis* L. *Journal of Horticultural Science* 63: 705-709.

Litz, R.E., Moore, G.A., and Srinivasan, C. (1985). In vitro systems for propagation and improvement of tropical fruits and palms. *Horticultural Review* 7: 157-200.

Murashige, T., Bitters, W.P., Rangan, E.M., Nauer, E.M., Roistacher, C.N., and Holliday, P.B. (1972). A technique of shoot apex grafting and its utilization towards recovering virus-free citrus clones. *HortScience* 7: 118-119.

Navarro, L., Roistacher, C.N., and Murashige, T. (1975). Improvement of shoot-tip grafting in vitro for virus-free citrus. *Journal of American Society for Horticultural Science* 100: 471-479.

## Chapter 12

Bernier, G. (1988). The control of floral evocation and morphogenesis. *Annual Review of Plant Physiology and Molecular Biology* 39: 175-219.

Bernier, G., Corbesier, L., Perilleux, C., Havelange, A., and Lejeune, P. (1998). Physiological analysis of the floral transition. In Cockshull, K. E., Gray, D., Seymour, G. B., Thomas, B. (Eds.) *Genetic and Environmental Manipulation of Horticultural Crops.* (pp. 103-110). Oxon, UK: CABI Publishing.

Bernier, G., Havelange, A., Houssa, C., Petitjean, A., and Lejeune, P. (1993). Physiological signals that induce flowering. *The Plant Cell* 5: 1147-1155.

Bernier, G., Kinet, M., and Sachs, R. M. (1981a). *The Physiology of Flowering. I. The Initiation of Flowers.* Boca Raton, FL: CRC Press.

Bernier, G., Kinet, M., and Sachs, R. M. (1981b). *The Physiology of Flowering. II. Transition to Reproductive Growth.* Boca Raton, FL: CRC Press.

Colasanti, J. and Sundaresan, V. (1996). Control of the transition to flowering. *Current Opinion in Biotechnology* 7: 145-149.

Murashige, T. and Skoog, F. (1962). A revised medium for rapid growth and bioassays with tobacco tissue cultures. *Physiologia Plantarum* 15: 473-497.

Nadgauda, R. S., Prasharami, V. A., and Mascarenhas, A. F. (1990). Precocious flowering and seeding behaviour in tissue cultured bamboos. *Nature* 344: 335-336.

Van Staden, J. and Dickens, C. W. S. (1991). In vitro induction of flowering and its relevance to micropropagation. In Bajaj, Y. P. S. (Ed.) *Biotechnology in Agriculture and Forestry, vol 17. High-Tech and Micropropagation* 1, (pp. 85-115). Berlin: Springer-Verlag.

Weigel, D. and Meyerowitz, E. (1994). The ABCs of floral homeotic genes. *Cell* 78:203-209.

Weigel, D. and Nilsson, O. (1995). The genetics of flower development. From floral induction to ovule morphogenesis. *Annual Review of Genetics* 29: 19-39.

## Chapter 13

Baker, W. G. (1953). A method for the in vitro culturing of potato tubers. *Science* 118: 384-385.

Harmey, M. A., Rowley, M. P., and Clinch, P. E. M. (1966). The effect of growth regulators on tuberisation of cultured stem pieces of *Solanum tuberosum*. *European Potato Journal* 9: 146-151.

Hussey, G. and Stacey, N. J. (1984). Factors affecting the formation of in vitro tubers of potato (*Solanum tuberosum* L.). *Annals of Botany* 53: 565-578.

Jean, M. and Cappadocia, M. (1991). In vitro tuberisation in *Dioscorea alata* L. 'Brazo fuerte' and 'Florido' and *D. abyssinica* Hoch. *Plant Cell Tissue and Organ Culture* 26: 147-152.

Murashige, T. and Skoog, F. (1962). A revised medium for rapid growth and boiassays with tobacco cultures *Physiologia Plantarum* 15: 473-497.

Seabrook, J. E. A., Coleman, S., and Levy, D. (1993). Effect of photoperiod on in vitro tuberisation of potato (*Solanum tuberosum* L.). *Plant Cell Tissue and Organ Culture* 34:43-51.

## Chapter 14

Cox, T. S. (1998). Deepening the wheat gene pool. *Journal of Crop Production* 1: 1-25.

Ellis, R. P., Forster, B. P., Robinson, D., Handley, L. L., Gordon D. C., Russell J. R., and Powell W. (2000). Wild barley: A source of genes for crop improvement in the 21st century? *Journal of Experimental Botany* 51: 9-17.

Eujayl, I., Baum, M., Powell, W., Erskine, W., and Pehu, E. (1998). A genetic linkage map of lentil (*Lens* sp.) based on RAPD and AFLP markers using recombinant inbred lines. *Theoretical and Applied Genetics* 97: 83-89.

Forster, B. P., Ellis, R. P., Newton, A. C., Tuberosa, R., This, D., El-Gamal, A. S., Bahri, M. H., and Ben Salem, M. (1999). Molecular breeding of barley for droughted low input agricultural conditions. In *Plant Nutrition: Molecular Biology and Genetics*. Gissel-Nielsen, G. and Jensen, A. (Eds.). (pp 359-363). Dordrecht: Kluwer Academic Publishers.

Karp, A., Isaac, P. G., and Ingram, D. (1998). *Molecular Tools for Screening Biodiversity*. New York: Chapman and Hall.

Loo, A. H. B., Tan, H. T. W., Kumar, P. P., and Saw, L. G. (1999). Population analysis of *Licuala glabra* Griff. var. *glabra* using RAPD profiling. *Annals of Botany* 84: 421-427.

Paul, S., Wachira, F. N., Powell, W., and Waugh, R. (1997). Diversity and genetic differentiation among populations of Indian and Kenyan tea (*Camellia sinensis* (L) O Kuntze) revealed by AFLP markers. *Theoretical and Applied Genetics* 94: 255-263.

Rajaseger, G., Tan, H. T. W., Turner, I. M., and Kumar, P. P. (1997). Analysis of genetic diversity among *Ixora* cultivars (Rubiaceae) using Random Amplified Polymorphic DNA. *Annals of Botany* 80: 355-361.

Saghai Maroof, M. A., Biyashev, R. M., Yang, G. P., Zhang, Q., and Allard, R. W. (1994). Extraordinarily polymorphic microsatellite DNA in barley: Species diversity, chromosomal locations, and population dynamics. *Proceedings of National Academy of Science USA* 91:5466-5470.

Swanston, J. S., Thomas, W. T. B., Powell, W., Young, G. R., Lawrence, P. E., Ramsay, L., and Waugh, R. (1999). Using molecular markers to determine barleys most suitable for malt whisky distilling. *Molecular Breeding* 5: 103-109.

Toojinda, T., Baird, E., Booth, A., Broers, L., Hayes, P., Powell, W., Thomas, W., Vivar, H., and Young, G. (1998). Introgression of quantitative trait loci (QTLs) determining stripe rust resistance in barley: An example of marker-assisted line development. *Theoretical and Applied Genetics* 96: 123-131.

# Index

Abscisic acid (ABA), 130
Accelerated backcrossing, 144-145,
    147
Adventitious caulogenesis, 24
Adventitious shoots, 22
*Agrobacterium*-mediated gene transfer,
    139
Ahloowalia, B. S., 94
Albinism, 55
Alkylating agents, 97
Allard, R. W., 141, 147
Amino acid accumulators, 88-90
Amplified fragmented length
    polymorphism (AFLP), 109,
    140, 141, 144, 145, 146
Androgenesis
    condition of donor plants, 48-50
    defined, 47
    favorable conditions, 51-53
    history of, 48
    importance of, 53
    regeneration of diploid plants, 53-54
    underlying principle, 51
Anther
    anatomy, 45-47
    culture, 47-50
    wall composition, 47
    wall derivation, 46-47
Antioxidants, 20
Apex grafting, 122-123
Asparagus, commercial
    micropropagation, 30
Auxins
    discovery of, 1
    flower inhibitor, 128-129
    growth in vitro, 8, 9, 11-12
    in pretransplant stage, 25-26
Axillary caulogenesis, 22, 23, 24

Backcrossing, 144-145, 147
Bahri, M. H., 146
Baird, E., 144
Bajaj, U. P. S., 111, 112, 119
Baker, W. G., 136
Balatková, V., 63
Bamboo, 133-134
Barley breeding, 146-147
Base analogs, 96-97
Baum, M., 146
Bean callus, 13
Belliard, G., 3
Ben Salem, M., 146
Bennetzen, J. F., 106
Bernier, G., 127, 129, 130
Bhojwani, S. S., 54, 103
Binder, B. M., 93
Bingham, E. T., 106
Biology, 4
Biotechnology, 4
Bitters, W. P., 121
Biyashev, R. M., 141, 147
Bleecker, A. B., 93
Booth, A., 144
Bourgin, J. P., 3, 48
Bowman, J. L., 99, 100
Breeding
    molecular. *See* Molecular breeding
    mutation, 37-44
Breznovitis, A., 86
Broers, L., 144
Broertjes, C., 41

Calcium in modification of auxin-
    cytokinin action, 13-14
Callogenesis, 21-22

*159*

T - #0194 - 101024 - C0 - 229/152/10 [12] - CB - 9781560229070 - Gloss Lamination